Why Us?

How Science Rediscovered the Mystery of Ourselves

JAMES LE FANU

Harper
Press

For Juliet

Harper*Press*
An imprint of HarperCollins*Publishers*
77–85 Fulham Palace Road,
Hammersmith, London W6 8JB
www.harpercollins.com

This Harper*Press* paperback edition published 2010
1

First published by Harper*Press* in 2009

A catalogue record for this book is available from the British Library

ISBN 978-0-00-712028-4

Set in Minion

Printed and bound in Great Britain by Clays Ltd, St Ives plc

Contents

Illustrations

Introduction: A Mystery to Ourselves

> 'Know then thyself, presume not God to scan;
> The proper study of mankind is man . . .
> Sole judge of Truth, in endless Error hurled:
> The glory, jest and *riddle* of the world!'
>
> *Alexander Pope, 'An Essay on Man' (1734)*

'Wonders are there many,' observed the Greek dramatist Sophocles – 'but none more wonderful than Man.' And rightly so for man, as far as we can tell, is the sole witness of the splendours of the universe he inhabits – though consistently less impressed by his existence than would seem warranted.

> 'Men go abroad to wonder at the height of mountains, at the huge waves of the sea, at the long courses of the rivers, at the vast compass of the ocean, at the circular motion of the stars,' observed St Augustine in the fifth century AD, '*and they pass by themselves without wondering.*'

The reasons for that lack of wonder at ourselves have changed over the centuries, but the most important still stands: the practicalities of our everyday lives are so simple and effortless as to seem unremarkable. We open our eyes on waking to be surrounded immediately by the shapes and colours, sounds, smells and movement of the world around us in the most vivid and exquisite detail. We feel hungry, and by some magical alchemy of which we know nothing, our bodies transform the food and drink before us into our flesh and blood. We open our mouths to speak and the words flow, a ceaseless babbling brook of thoughts and ideas and impressions. We reproduce our kind with ease and play no part in the transformation, in three short months, of the single fertilised egg into an embryo, no larger than a thumbnail, whose four thousand functioning parts include a beating heart the size of the letters on this page, and a couple of eyes the size of the full stop at the end of this sentence. We attend to our children's needs, but effortlessly they grow inch by inch, year by year to adulthood, replacing along the way virtually every cell in their bodies, refashioning the skull, limbs and internal organs, while retaining the proportions of one part to another.

The moment one starts to reflect on any of these practicalities, their effortlessness does begin to seem rather astonishing. They clearly are not in the least bit simple – yet in reality they are almost the simplest thing we know. They appear simple *because they have to be so*: if our senses did not accurately capture the world around us, if our metabolism did not abstract and utilise every nutrient, if procreation was not almost too easy and the growth of children into adulthood not virtually automatic, if we had to consciously make an effort to speak a sentence – then 'we' would never have happened.

This should make us pause for a moment because, from common experience, there is nothing more difficult and arduous than to make the complex appear simple – just as the concert pianist's seemingly effortless keyboard skills are grounded in years of toil and practice. So, it is precisely the *effortlessness* of our everyday lives that should command our attention – recognising their semblance of simplicity as a mark of their profundity.

But most people nowadays do 'pass by themselves without wondering'; though less justifiably so than in St Augustine's time, for we now know prodigiously more about the deep biological complexities that underpin those simplicities of our everyday lives. We should, by rights, be enormously *more* appreciative of nature's ingenuity, and the deceptive effortlessness of our seeing and talking and reproducing our kind should be part of common knowledge, a central theme of the school biology curriculum, promoting a sense of wonder in the young mind at the fact of its very existence.

Yet one could search a shelf full of biology text books in vain for the slightest hint of the extraordinary in their detailed exposition of these practicalities of our lives. And why? Scientists do not 'do' wonder. Rather, for the past 150 years they have interpreted the world through the prism of supposing that there is nothing in principle that cannot be accounted for, where the unknown is merely waiting-to-be-known. And so it has been till very recently, when two of the most ambitious scientific projects ever conceived have revealed, quite unexpectedly – and without anyone really noticing – that we are after all a mystery to ourselves. This is the story of how it happened, and its (many) consequences.

1

Science Triumphant, Almost

> 'The real voyage of discovery consists not in seeking new lands, but in seeing with new eyes.'
>
> *Marcel Proust*

We live in the Age of Science, whose lengthy roll-call of discoveries and technical innovations has immeasurably changed our lives for the better. Within living memory children succumbed in their thousands every year from polio and whooping cough, telephones were a rarity, colour television was yet to be invented and the family would gather every evening around the wireless after supper to listen to the news.

Since then, the therapeutic revolution of the post-war years has reduced death in infancy to its irreducible minimum, while ensuring that most now live out their natural lifespan; the electronic revolution has prodigiously extended both the capacity of the human mind, with computers of ever smaller size and greater power, and its horizons, with the Hubble telescope circling in orbit around the earth, relaying back from the far reaches of the cosmos sensational images of its beauty and grandeur.

The landmarks of this post-war scientific achievement are familiar enough: for medicine, there are antibiotics and the pill, heart transplants and test tube babies (and much else besides); for electronics, the mobile phone and the Internet; for space exploration, the Apollo moon landing of 1969 and the epic journey of Voyagers I and II to the far reaches of our solar system. But these last fifty years have witnessed something yet more remarkable still – a series of discoveries that, combined together, constitute the single most impressive intellectual achievement of all time, allowing us to 'hold in our mind's eye' the entire sweep of the history of the universe from its beginning till now. That history, we now know, starts fifteen thousand million years ago (or thereabouts) with the Big Bang, 'a moment of glory too swift and expansive for any form of words [when] a speck of matter became in a million millionth of a second something at least ten million million million times bigger'. Eleven thousand million years pass, and a massive cloud of gas, dust, pebbles and rocks in a minor galaxy of that (by now) vast universe coalesces around a young sun to create the planets of our solar system. Another thousand million years pass, the surface of the earth cools and the first forms of life emerge from some primeval swamp of chemicals. Yet another two and a half thousand million years elapse till that moment a mere(!) five million years ago when the earliest of our ancestors first walked upright across the savannah plains of central Africa.

And again, within living memory we knew none of this, neither how the universe came into being, nor its size and composition; neither how our earth was born, nor how its landscape and oceans were created; neither the timing of the emergence of life, nor the 'universal code' by which all living things reproduce their kind; neither the physical characteristics of our earliest ancestors, nor the details of their evolutionary transformation to modern man. Now we do, and holding this historical sweep 'in our mind's eye' it is possible to appreciate the intellectual endeavour that underpins it will never, *can* never, be surpassed. How astonishing to realise that today's astronomers can detect the distant echoes of that 'moment of glory' of the Big Bang all those billions of years ago, and capture in those astonishing images transmitted from the Hubble telescope the very processes that brought our solar system into existence. How astonishing that geologists should have discovered that massive plates of rock beneath the earth's surface, moving at the rate of a centimetre a year, should have formed its continents and oceans, the mountains and valleys of the snow-capped Himalayas thrust upwards by the collision of the Indian subcontinent with the Asian landmass. How astonishing, too, that biologists should now understand the internal workings of the microscopic cell, and how the arrangements of the same four molecules strung out along the elegant spiral of the Double Helix contain the 'master plan' of every living thing that has ever existed.

It is impossible to convey the intellectual exhilaration of such momentous discoveries, but the account by Donald Johanson of finding the first near-complete skeleton of our three-and-a-half-million-year-old hominid ancestor 'Lucy' conveys something of the emotions felt by so many scientists over the past fifty years.

> Tom [Gray] and I had surveyed for a couple of hours. It was now close to noon, and the temperature was approaching 110. We hadn't found much: a few teeth of a small extinct horse; part of the skull of an extinct pig, some antelope molars, a bit of a monkey jaw . . .
>
> 'I've had it,' said Tom. 'When do we head back to camp?'
>
> But as we turned to leave, I noticed something lying on the ground part way up the slope.
>
> 'That's a bit of a hominid arm,' I said.
>
> 'Can't be. It's too small. Has to be monkey of some kind.'
>
> We knelt to examine it.
>
> 'Much too small,' said Gray again.
>
> I shook my head. 'Hominid.'
>
> 'What makes you so sure?' he said.
>
> 'That piece right next to your hand. That's hominid too.'
>
> 'Jesus Christ,' said Gray. He picked it up. It was the back of a small skull. A few feet away was part of a femur; a thigh bone. 'Jesus Christ,' he said again. We stood up and began to see other bits of bone on the slope. A couple of vertebrae, part of a pelvis – all of them hominid. An unbelievable, impermissible thought flickered through my mind. For suppose all these fitted together? Could they be parts of a single extremely primitive skeleton? No such skeleton has ever been found – anywhere.
>
> 'Look at that,' said Gray. 'Ribs.'
>
> A single individual.
>
> 'I can't believe it,' I said, 'I just can't believe it.'
>
> 'By God you'd better believe it!' shouted Gray. His voice went up into a howl. I joined him. In that 110 degree heat we began jumping up and down. With nobody to share our feelings, we hugged each other, sweaty and smelly, howling and hugging in the heat-shimmering gravel, the small brown remains of what now seemed almost certain to be parts of a single hominid skeleton lying all around us.

Momentous events have multiple causes, and the source of this so recent and all-encompassing delineation of the history of our universe stretches back across the centuries. It is impossible to hope to convey the intellectual brilliance and industry of those who brought this extraordinary enterprise to fruition, whose major landmarks are summarised here as the Thirty Definitive Moments of the past six decades.

TABLE 1

Science Triumphant 1945–2001: Thirty Definitive Moments

1945	The atom bomb: Hiroshima and Nagasaki
1946	The electron microscope reveals the internal structure of the cell
1947	The invention of the transistor launches the Electronic Age
1953	Theory of formation of the chemical elements of life by nuclear fusion within stars
1953	The laboratory simulation of the 'origin of life'
1953	James Watson and Francis Crick discover the Double Helix
1955	The first polio vaccine
1957	The Soviet Union launches Sputnik and the epoch of planetary exploration
1960	The oral contraceptive
1961	The Genetic Code deciphered
1965	The theory of the Big Bang confirmed by discovery of cosmic microwave background radiation
1967	The first heart transplant
1969	US astronaut Neil Armstrong becomes the first man on the moon
1969	James Lovelock proposes theory of a life-sustaining atmosphere
1973	The advent of genetic engineering
1973	The invention of magnetic resonance imaging of the brain
1974	The discovery of 'Lucy', *Australopithecus afarensis*, dated 4 million years BC
1974	The first Grand Unified Theory of particle physics
1977	The first complete genetic sequence of an organism
1977	The first personal computer designed for the mass market
1979	Voyagers I and II relay data from Jupiter, Saturn, Uranus and Neptune

1979 The first 'test tube baby'
1980 The asteroid impact hypothesis of the mass extinction of dinosaurs
1984 The discovery of 'Turkana Boy', the first complete skeleton of *Homo erectus*, dated 1.5 million years BC
1984 Confirmation of theory of plate tectonics
1987 Formulation of the 'out of Africa' hypothesis of human evolution
1989 Launch of world wide web
1990 The Decade of the Brain
1999 The Hubble space telescope observes the birth of stars in the constellation Taurus
2001 Publication of the Human Genome

The triumph of science, one might suppose, is virtually complete. What, during these times, have we learned from the humanities – philosophy, say, theology or history – that begins to touch the breadth and originality of this scientific achievement and the sheer extraordinariness of its insights? What, one might add, have the humanities done that begins to touch the medical therapeutic revolution of the post-war years or the wonders of modern technology?

That history of our universe as revealed in the recent past draws on many disciplines: cosmology and astronomy obviously, the earth and atmospheric sciences, biology, chemistry and genetics, anthropology and archaeology, and many others. But science is also a unified enterprise, and these areas of enquiry all 'hang together' to reveal the coherent story outlined above. There remained, however, two great unknowns, two final obstacles to a truly comprehensive theory that would also explain our place in that universe.

The first is how it is that we, like all living things, reproduce our kind with such precision from one generation to the next. The 'instructions', as is well recognised, come in the form of genes strung out along the two intertwining strands of the Double Helix in the nucleus of every cell. But the question still remained: How do those genes generate that near-infinite diversity and beauty of form, shape and size, and behaviour that distinguish one form of life from another? How do they fashion from a single fertilised human egg the unique physical features and mind of each one of us?

The second of these 'great unknowns' concerned the workings of the brain, and the human brain in particular. To be sure, neurologists have over the past hundred years identified the functions of its several parts – with the frontal lobes as the 'centre' of rational thought and emotion, the visual cortex at the back, the speech centre in the left hemisphere and so on. But again the question remained: How *does* the electrical firing of the brain's billions of nerves 'translate' into our perception of the sights and sounds of the world around us, our thoughts and emotions and the rich inner landscape of personal memories?

These two substantial questions had remained unresolved because both the Double Helix and the brain were inaccessible to scientific scrutiny: the Double Helix, with its prodigious amount of genetic information, comes packed within the nucleus of the cell, a mere one five thousandth of a millimetre in diameter; while the blizzard of electrical activity of the billions of neurons of the brain is hidden within the confines of the bony vault of the skull. But then, in the early 1970s, a series of technical innovations would open up first the Double Helix and then the brain to scientific investigation, with the promise that these final obstacles to our scientific understanding of ourselves might soon be overcome. We will briefly consider each in turn.

The Double Helix

The Double Helix, discovered by James Watson and Francis Crick in 1953, is among the most familiar images of twentieth-century science. Its simple and elegant spiral structure of two intertwined strands unzips and replicates itself every time the cell divides – each strand, an immensely long sequence of just four molecules (best conceived, for the moment, as four different-coloured discs – blue, yellow, red and green). The specific arrangement of a thousand or more of these coloured discs constitutes a 'gene', passed down from generation to generation, that determines your size and shape, the colour of your eyes or hair or any other similarly distinguishing traits, along with the thousands of widgets or parts from which we are all made. It would take another fifteen years to work it all out, at least in theory – but the practical details of which particular sequence of coloured discs constituted which gene, and what each gene *did*, still remained quite unknown. This situation would change dramatically in the 1970s, with three technical innovations that would allow biologists first to chop up those three billion 'coloured discs' into manageable fragments, then to generate thousands of copies the better to study them, and finally to 'spell out' the sequence (red, green, blue, green, yellow, etc., etc.) that constitutes a single gene.

Fig 1-1: *James Watson (left) and Francis Crick (right) in the Cavendish laboratory in Cambridge point out the main features of their first full model of the Double Helix.*

It lies beyond hyperbole to even try to convey the excitement and exhilaration generated by this trio of technical innovations, whose potential marked 'so significant a departure from that which had gone before' they would become known collectively as 'the New Genetics'. The prospect of deciphering the genetic instructions of 'life' opened up a Pandora's box of possibilities, conferring on biologists the opportunity to change the previously immutable laws of nature by genetically modifying plants and animals. The findings of the New Genetics filled the pages not only of learned journals but of the popular press: 'Gene Find Gives Insight into Brittle Bones', 'Scientists Find Genes to Combat Cancer', 'Scientists Find Secret of Ageing', 'Gene Therapy Offers Hope to Victims of Arthritis', 'Cell Growth Gene Offers Prospect of Cancer Cure', 'Gene Transplants to Fight Anaemia', and so on.

The New Genetics, in short, swept all before it to become synonymous with biology itself. Before long the entire spectrum of research scientists – botanists, zoologists, physiologists, microbiologists – would be applying its techniques to their speciality. The procedures themselves in turn became ever more sophisticated, opening up the prospect that the New Genetics might transcend the possibilities of discovering 'the gene for this and the gene for that', to spell out the *entire* sequence of coloured discs strung along the Double Helix and thus reveal the full complement of genes, known as The Genome. There was every reason to suppose that deciphering the full set of genetic instructions of what makes a bacterium a bacterium, a worm a worm, and a common housefly a common housefly would reveal how they are made and how they come to be so readily distinguishable from each other – why the worm should burrow and the fly should fly. Then, at the close of the 1980s, the co-discoverer of the Double Helix, James Watson, proposed what would become the single most ambitious and costly project ever conceived in the history of biology – to spell out the full complement of human genes. Thus the Human Genome Project (HGP) was born, with its promise to make clear what it is in our genes that makes us, 'us'. The truism that 'the answer lies in the genes' is not merely an abstract idea, rather the set of instructions passed down from generation to generation influences every aspect of our being: our physical characteristics, personality, intelligence, predisposition to alcoholism or heart disease, and much else besides. Spell out the human genome in its entirety, and all these phenomena, and more, should finally be accounted for.

'The search for this "Holy Grail" of who we are,' observed Harvard University's Walter Gilbert, 'has now reached its culminating phase, the ultimate goal is the acquisition of all the details of our genome . . . that will transform our capacity to predict what we will become.' There could be no greater aspiration than to 'permit a living creature', as Robert Sinsheimer, Chancellor of the University of California, put it, 'for the first time *in all time*, to understand its origins and design its future'. The Genome Project would, claimed Professor John Savile of Nottingham's University Hospital, 'like a mechanical army, systematically destroy ignorance', while 'promising unprecedented opportunities for science and medicine'.

The Human Genome Project was formally launched in 1991, with a projected cost of $3 billion over the fifteen years it was expected it would run. The task of assembling the vast quantity of data generated by spelling out the human genome was divided across several centres, most of them in the United States, Britain and Japan. The scene within those 'genomes centres' could have come from a science fiction film – gleaming automated machines as far as the eye could see, labelling each chemical of the Double Helix with its own fluorescent dye which was then 'excited' by a laser and the results fed directly into a computer. 'The Future is Now' trumpeted the cover of *Time* magazine, imposing that iconic image of the Double Helix over the shadowy outline of a human figure in the background.

The Brain

Meanwhile, the human brain too was about to reveal its secrets. Its physical appearance is quite as familiar as the Double Helix. But the specialisation of those separate parts for seeing, hearing, movement and so on is in a sense deceptive, concealing the crucial question of *how* their electrical firing translates the sights and sounds of the external world, or summons those evocative childhood memories from the distant past. How does this mere three pounds of soft grey matter within the skull contain the experience of a lifetime?

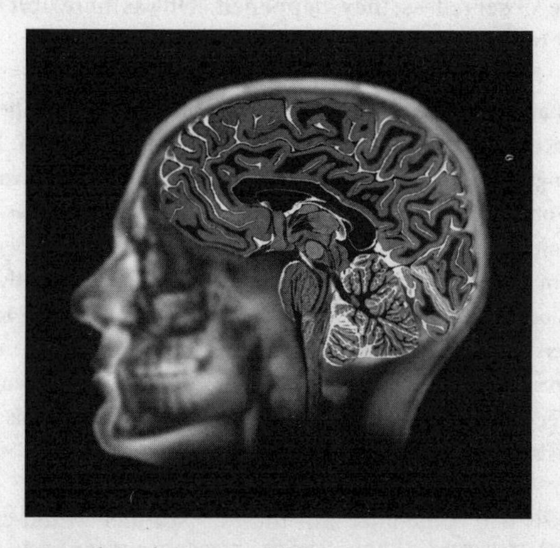

Fig 1-2: *From the early 1970s, sophisticated scanning techniques would permit scientists for the first time to scrutinise the intricate structure of the living brain.*

Here again, a series of technical innovations, paralleling those of the New Genetics, would permit scientists for the first time to scrutinise the brain 'in action'. In 1973 the British physicist Godfrey Hounsfield invented the Computed Tomography (CT) scanner, revealing the brain's internal structure with an almost haunting clarity, revolutionising the diagnosis of strokes and tumours and other forms of mischief. Soon after, the further technical development of Positron Emission Tomography (PET) scanning would transform the CT scanner's static images or 'snapshots' of the brain into 'moving pictures'.

Put simply, this is how it works. All of life requires oxygen to drive the chemical reactions in its cells. This oxygen is extracted from the air, inspired in the lungs and transported by blood cells to the tissues. When, for example, we start talking, the firing of the neurons in the language centre of the brain massively increases their demand for oxygen, which can only be met by increasing the bloodflow to that area. The PET scanner detects that increase in bloodflow, converting it into multi-coloured images that pick out the 'hotspots' of activity. Now, for the first time, the internal workings of the brain induced by smelling a rose or listening to a violin sonata could be observed as they happened. Or (as here) picking out rhyming words:

> A woman sits quietly waiting for the experiment to begin – her head ensconced in a donut-shaped device, a PET scanning camera. Thirty-one rings of radiation detectors make up the donut, which will scan thirty-one images simultaneously in parallel horizontal lines. She is next injected with a radioactive isotope [of oxygen] and begins to perform the task . . . The words are presented one above the other on a television monitor. If they rhyme, she taps a response key. Radiation counters estimate how hard the brain region is working . . . and are transformed into images where higher counts are represented by brighter colours [thus] this colour map of her brain reveals all the regions acting while she is judging the paired words.

The details will come later, but the PET scanner would create the discipline of modern neuroscience, attracting thousands of young scientists keen to investigate this previously unexplored territory. Recognising the possibilities of the new techniques, the United

States Congress in 1989 designated the next ten years as 'the Decade of the Brain' in anticipation of the many important new discoveries that would deliver 'precise and effective means of predicting, modifying and controlling individual behaviour'. 'The question is not whether the neural machinery [of the brain] will be understood,' observed Professor of Neurology Antonio Damasio, writing in the journal *Scientific American*, 'but when.'

Throughout the 1990s, both the Human Genome Project and the Decade of the Brain would generate an enormous sense of optimism, rounding off the already prodigious scientific achievements of the previous fifty years. And sure enough, the completion of both projects on the cusp of the new millennium would prove momentous events.

The completion of the first draft of the Human Genome Project in June 2000 was considered sufficiently important to warrant a press conference in the presidential office of the White House. 'Nearly two centuries ago in this room, on this floor, Thomas Jefferson spread out a magnificent map . . . the product of a courageous expedition across the American frontier all the way to the Pacific,' President Bill Clinton declared. 'But today the world is joining us here to behold a map of even greater significance. We are here to celebrate the completion of the first survey of the entire human genome. Without a doubt this is the most important, most wondrous map ever produced by mankind.'

The following year, in February 2001, the two most prestigious science journals, *Nature* and *Science*, each published a complete version of that 'most wondrous map ever produced by mankind' as a large, multi-coloured poster displaying the full complement of (as it would turn out) twenty-five thousand human genes. It was, as *Science* observed, 'an awe-inspiring sight'. Indeed, it was awesome twice over. Back in the 1950s, when Francis Crick and James Watson were working out the structure of the Double Helix, they had no detailed knowledge of a *single* gene, what it is or what it does. Now, thanks to the techniques of the New Genetics, those involved in the Genome Project had, in less than a decade, successfully culled from those three billion 'coloured discs' strung out along its intertwining strands the hard currency of each of the twenty-six thousand genes that determine who we are.

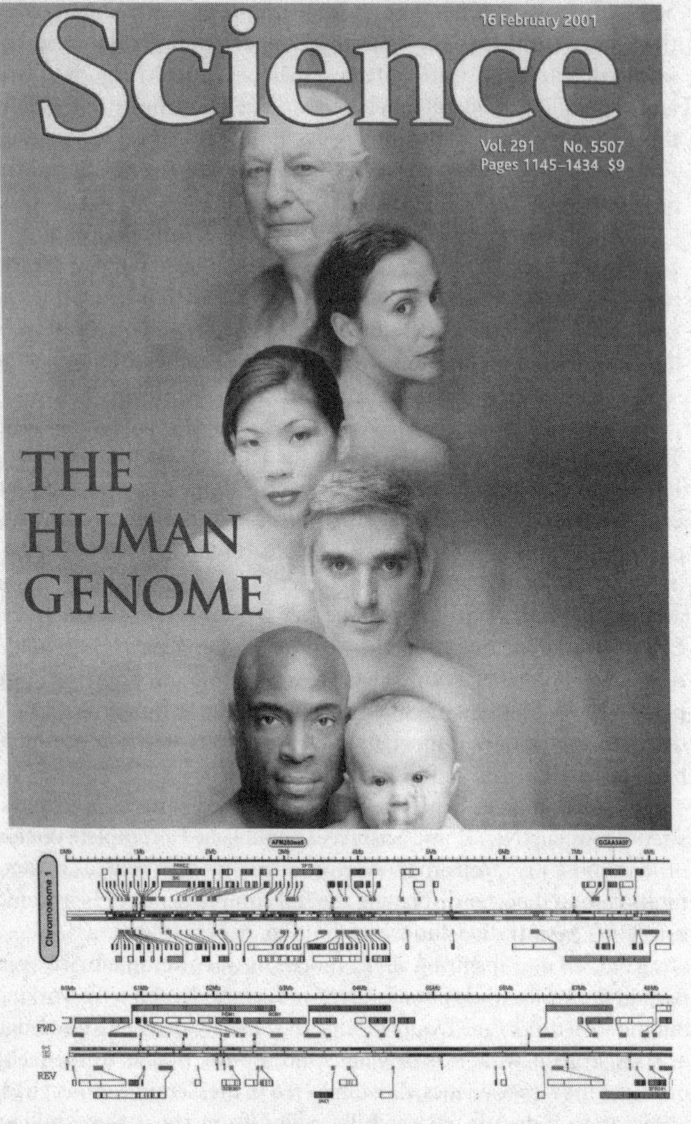

Fig 1-3: *The front cover of the Human Genome issue of* Science, *featuring 'the most wondrous map ever produced by mankind'. Below, the opening sequence of chromosome 1 reveals the highly technical nature of the genetic information.*

The Human Genome map, like Thomas Jefferson's map of the United States, portrays the major features of that genetic landscape with astonishing precision. While it had taken the best part of seven years to find the defective gene responsible for the lung disorder cystic fibrosis, now anyone could locate it from that multi-coloured poster in as many seconds. Here too at a glance you can pick out the gene for the hormone insulin, which controls the level of sugar in the blood, or the haemoglobin molecule that transports oxygen to the tissues. To be sure, the functions of many thousands of those genes remained obscure, but now, knowing their precise location and the sequence of which they are composed, it would be only a matter of time before they too would be known. It was a defining moment. 'Today will be recorded as one of the most significant dates in history,' insisted one of the major architects of the Genome Project, Dr Michael Dexter of the Wellcome Trust in Britain. 'Just as Copernicus changed our understanding of the solar system and man's place within it, so knowledge of the human genome will change how we see ourselves and our relationship to others.'

The goals of the Decade of the Brain were necessarily more open-ended, but still the PET scanner, and the yet more sophisticated brain imaging techniques that followed in its wake, had more than fulfilled their promise, allowing scientists to draw another exquisitely detailed map locating the full range of mental abilities to specific parts of the brain. There were many surprises along the way, not least how the brain fragmented the simplest of tasks into a myriad of different components. It had long been supposed, for instance, that the visual cortex at the back of the brain acted as a sort of photographic plate, capturing an image of the external world as seen through the eye. But now it turned out that the brain 'created' that image from the interaction of thirty or more separate maps within the visual cortex, each dedicated to one or other aspect of the visual image, the shapes, colour, movement of the world 'out there'. 'As surely as the old system was rooted in the concept of an image of the visual world received and analysed by the cortex,' observes Semir Zeki, Professor of Neurobiology at the University of London, 'the present one is rooted in the belief that an image of the visual world is *actively* constructed by the cerebral cortex.'

Steven Pinker, Professor of Brain and Cognitive Science at the Massachusetts Institute of Technology, could explain to the readers of *Time* magazine in April 2000 (the close of the Decade of the Brain) how neuroscientists armed with their new techniques had investigated 'every facet of mind from mental images to moral sense, from mundane memories to acts of genius', concluding, 'I have little reason to doubt that we will crack the mystery of how brain events correlate with experience.'

Both the Human Genome Project and the Decade of the Brain have indeed transformed, beyond measure, our understanding of ourselves – but in a way quite contrary to that anticipated.

Nearly ten years have elapsed since those heady days when the 'Holy Grail' of the scientific enterprise, the secrets of life and the human mind, seemed almost within reach. Every month the pages of the science journals are still filled with the latest discoveries generated by the techniques of the New Genetics, and yet more colourful scans of the workings of the brain – but there is no longer the expectation that the accumulation of yet more facts will ever provide an adequate scientific explanation of the human experience. Why?

We return first to the Human Genome Project, which, together with those of the worm and fly, mouse and chimpanzee and others that would follow in its wake, was predicated on the assumption that knowledge of the full complement of genes must explain, to a greater or lesser extent, why and how the millions of species with which we share this planet are so readily distinguishable in form and attributes from each other. The genomes must, in short, reflect the complexity and variety of 'life' itself. But that is not how it has turned out.

First, there is the 'numbers problem'. That final tally of twenty-five thousand human genes is, by definition, sufficient for its task, but it seems a trifling number to 'instruct', for example, how a single fertilised egg is transformed in a few short months into a fully formed being, or to determine how the billions of neurons in the brain are wired together so as to encompass the experiences of a lifetime. Those twenty-five thousand genes must, in short, 'multi-task', each performing numerous different functions, combining together in a staggeringly large number of different permutations.

That paucity of genes is more puzzling still when the comparison is made with the genomes of other creatures vastly simpler than ourselves – several thousand for a single-cell bacterium, seventeen thousand for a millimetre-sized worm, and a similar number for a fly. This rough equivalence in the number of genes across so vast a range of 'organismic complexity' is totally inexplicable. But no more so than the discovery that the human genome is virtually interchangeable with that of our fellow vertebrates such as the mouse and chimpanzee – to the tune of 98 per cent or more. There is, in short, *nothing* to account for those very special attributes that so readily distinguish us from our primate cousins – our upright stance, our powers of reason and imagination, and the faculty of language.

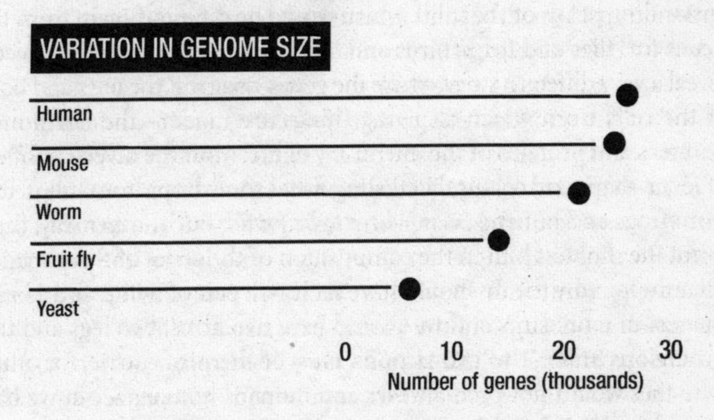

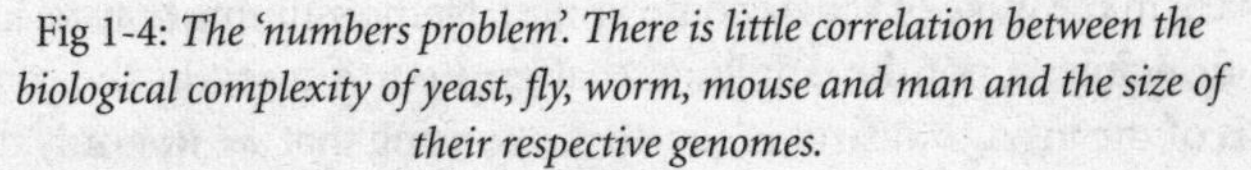

Fig 1-4: *The 'numbers problem'. There is little correlation between the biological complexity of yeast, fly, worm, mouse and man and the size of their respective genomes.*

The director of the Chimpanzee Genome Project, Svante Paabo, had originally anticipated that its comparison with the human genome would reveal the 'profoundly interesting genetic prerequisites' that set us apart:

> The realisation that a few genetic accidents made human history possible will provide us with a whole new set of philosophical challenges to think about . . . both a source of humility and a blow to the idea of human uniqueness.

But publication of the completed version of the chimpanzee genome in 2005 prompted a more muted interpretation of its significance: 'We cannot see in this why we are so different from chimpanzees,' Paabo commented. 'Part of the secret is hidden in there, but we don't understand it yet.' So 'The obvious differences between humans and chimps cannot be explained by genetics alone' – which would seem fair comment, until one reflects that if those differences 'cannot be explained' by genes, then what *is* the explanation?

These findings were not just unexpected, they undermined the central premise of biology: that the near-infinite diversity of form and attributes that so definitively distinguish living things one from the other must 'lie in the genes'. The genome projects were predicated on the assumption that the 'genes for' the delicate, stooping head and pure white petals of the snowdrop would be different from the 'genes for' the colourful, upstanding petals of the tulip, which would be different again from the 'genes for' flies and frogs, birds and humans. But the genome projects reveal a very different story, where the genes 'code for' the nuts and bolts of the cells from which all living things are made – the hormones, enzymes and proteins of the 'chemistry of life' – but the diverse subtlety of form, shape and colour that distinguishes snowdrops from tulips, flies from frogs and humans, is *nowhere to be found*. Put another way, there is not the slightest hint in the composition of the genes of fly or man to account for why the fly should have six legs, a pair of wings and a brain the size of a full stop, and we should have two arms, two legs and that prodigious brain. The 'instructions' *must* be there, of course, for otherwise flies would not produce flies and humans humans – but we have moved, in the wake of the Genome Project, from assuming that we knew the principle, if not the details, of that greatest of marvels, the genetic basis of the infinite variety of life, to recognising that we not only don't understand the principles, we have no conception of what they might be.

We have here, as the historian of science Evelyn Fox Keller puts it:

> One of those rare and wonderful moments when success teaches us humility . . . We lulled ourselves into believing that in discovering the basis for genetic information we had found 'the secret of life'; we were confident that if we could only decode the message in the sequence of chemicals, we would understand the 'programme' that makes an organism what it is. But now there is at least a tacit acknowledgement of how large that gap between genetic 'information' and biological meaning really is.

And so, too, the Decade of the Brain. The PET scanner, as anticipated, generated many novel insights into the patterns of electrical activity of the brain as it looks out on the world 'out there', interprets the grammar and syntax of language, recalls past events, and much else besides. But at every turn the neuroscientists found themselves completely frustrated in their attempts to get at *how the brain actually works.*

Right from the beginning it was clear that there was simply 'too much going on'. There could be no simpler experiment than to scan the brain of a subject when first reading, then speaking, then listening to, a single word such as 'chair'. This should, it was anticipated, show the relevant part of the brain 'lighting up' – the visual cortex when reading, the speech centre when speaking, and the hearing cortex when listening. But no, the brain scan showed that each separate task 'lit up' not just the relevant part of the brain, but generated a blizzard of electrical activity across vast networks of millions of neurons – while *thinking* about the meaning of a word and speaking appeared to activate the brain virtually in its entirety. The brain, it seemed, must work in a way previously never really appreciated – not as an aggregate of distinct specialised parts, but as an integrated whole, with the same neuronal circuits performing many different functions.

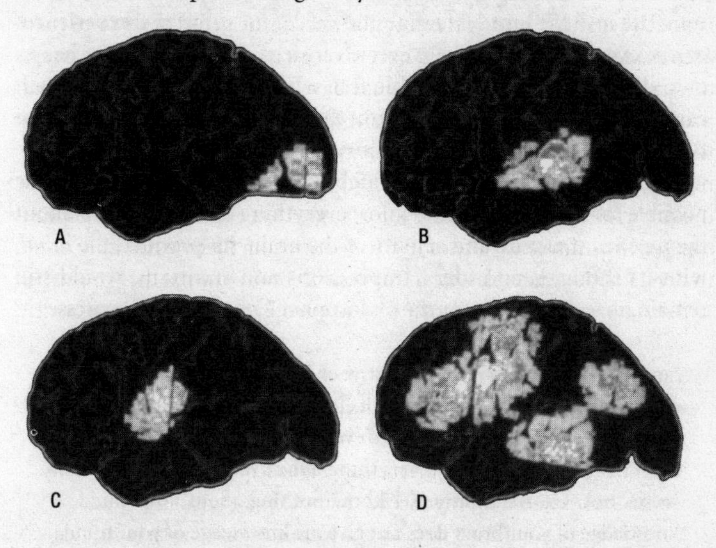

Fig 1-5: *'Too much going on': the simplest of intellectual tasks generates widespread electrical activity involving millions of neurons in the visual cortex when reading (A), the auditory cortex when listening (B) or the temporal lobe when thinking about words (C), while thinking and speaking (D) lights up vast tracts across the entire brain simultaneously.*

The initial surprise at discovering how the brain fragmented the sights and sounds of the world 'out there' into a myriad of separate components grew greater still as it became clear that there was no compensating mechanism that might reintegrate all those fragments back together again into that personal experience of being at the centre, moment by moment, of a coherent, ever-changing world. Reflecting on this problem of how to 'bind' all the fragments back together again, Nobel Prize-winner David Hubel of Harvard University observed:

> This abiding tendency for attributes such as form, colour and movement to be handled by separate structures in the brain immediately raises the question how all the information is finally assembled, say for perceiving a bouncing red ball. It obviously must be assembled – but where and how, *we have no idea.*

But the greatest perplexity of all was the failure to account for how the monotonous electrical activity of those billions of neurons in the brain translate into the limitless range and quality of subjective experiences of our everyday lives – where every transient, fleeting moment has its own distinct, unique, intangible *feel*: where the cadences of a Bach cantata are so utterly different from the flash of lightning, the taste of Bourbon from the lingering memory of that first kiss.

The implications are clear enough. While theoretically it might be possible for neuroscientists to know everything there is to know about the *physical* structure and activity of the brain, its 'product', the *mind*, with its thoughts and ideas, impressions and emotions, would still remain unaccounted for. As the philosopher Colin McGinn expresses it:

> Suppose I know everything about your brain: I know its anatomy, its chemical ingredients, the pattern of electrical activity in its various segments, I even know the position of every atom and its subatomic structure. Do I therefore know everything about your mind? It certainly seems not. On the contrary, I know nothing about your mind. So knowledge of your brain does not give me knowledge of your mind.

This distinction between the electrical activity of the *material* brain and the *non-material* mind (of thoughts and ideas) as two quite different things might seem so self-evident as to be scarcely worth commenting on. But for neuroscientists the question of how the brain's electrical activity translates into thoughts and sensations was precisely what needed explaining – and their failure to do so has come to haunt them. So, for everything that the Decade of the Brain undoubtedly achieved, nonetheless, as John Maddox, editor of *Nature*, would acknowledge at its close: 'We seem as far from understanding [the brain] as we were a century ago. Nobody understands how decisions are made or how imagination is set free.'

This verdict on the disappointing outcomes of the Genome Project and the Decade of the Brain might seem a trifle premature. These are, after all, still very early days, and it is far too soon to predict what might emerge over the next twenty to thirty years. The only certainty about advances in human knowledge is that they open the door to further seemingly unanswerable questions, which in time will be resolved, and so on. The implication that here science may finally have 'reached its limits' would seem highly contentious, having been expressed many times in the past, only to be repeatedly disproved. Famously, the physicist Lord Kelvin, at the close of the nineteenth century, insisted that the future of his discipline was to be looked for in 'the sixth place of decimals' (that is, futile refinements of the then present state of knowledge). Within a few years Albert Einstein had put forward his General Theory of Relativity, and the certainties of Lord Kelvin's classical physics were eclipsed.

The situation here, however, is rather different, for while the New Genetics and those novel brain scanning techniques offer almost inexhaustible opportunities for further research, it is possible to anticipate in broad outline what their findings will add up to. Scientists could, if they so wished, spell out the genomes of each of the millions of species with which we share this planet – snails, bats, whales, elephants and so on – but that would only confirm that they are composed of several thousand similar genes that 'code' for the nuts and bolts of the cells of which they are made, while the really interesting question, of how those genes determine the unique form and attributes of the snail, bat, elephant, whale or whatever, would remain unresolved. And so too for the scanning techniques of the neurosciences, where a million scans of subjects watching a video of bouncing red balls would not take us an iota further in understanding what needs explaining – how the neuronal circuits *experience* the ball as being red and round and bouncing.

At any other time these twin setbacks to the scientific enterprise might simply have been relegated to the category of problems for which science does not as yet have the answer. But when cosmologists can reliably infer what happened in the first few minutes of the birth of the universe, and geologists can measure the movements of vast continents to the nearest centimetre, then the inscrutability of those genetic instructions that should distinguish a human from a fly, or the failure to account for something as elementary as how we recall a telephone number, throws into sharp relief the unfathomability of ourselves. It is as if we, and indeed all living things, are in some way different, profounder and more complex than the physical world to which we belong.

Nonetheless there must be a reason why those genome projects proved so uninformative about the form and attributes of living things, or why the Decade of the Brain should have fallen so far short of explaining the mind. There is a powerful impression that science has been looking in the wrong place, seeking to resolve questions whose answers lie somehow outside its domain. This is not just a matter of science not yet knowing all the facts; rather there is the sense that something of immense importance is 'missing' that might transform the bare bones of genes into the wondrous diversity of the living world, and the monotonous electrical firing of the neurons of the brain into the vast spectrum of sensations and ideas of the human mind. What might that 'missing' element be?

Much of the prestige of science lies in its ability to link together disparate observations to reveal the processes that underpin them. But this does not mean that science 'captures' the phenomena it describes – far from it. There is, after all, nothing in the chemistry of water (two atoms of hydrogen to one of oxygen) that captures its diverse properties as we know them to be from personal experience: the warmth and wetness of summer rain, the purity and coldness of snow in winter, the babbling brook and the placid lake, water refreshing the dry earth, causing the flowers to bloom and cleansing everything it touches. It is customary to portray this distinction as 'two orders of reality'. The 'first' or 'primary reality' of water is that personal knowledge of its diverse states and properties that includes not just how we perceive it through our senses, but also the memories, emotions and feelings with which we respond to it. By contrast, the 'second order reality' is water's materiality, its chemical composition as revealed by the experimental methods of the founder of modern chemistry, the French genius Antoine

Lavoisier, who in 1783 sparked the two gases of hydrogen and oxygen together in a test tube, to find a residue of dew-like drops that 'seemed like water'.

These two radically different, yet complementary, 'orders of reality' of water are mutually exclusive. There is nothing in our personal experience that hints at water's chemical composition, nor conversely is there anything in its chemical formula that hints at its many diverse states of rain, snow, babbling brook, as we know them from personal experience. This seemingly unbridgeable gap between these two orders of reality corresponds, if not precisely, to the notion of the 'dual nature of reality', composed of a *non-material* realm, epitomised by the thoughts and perceptions of the mind, and an objective *material* realm of, for example, chairs and tables. They correspond, again if not precisely, to two categories of knowledge that one might describe respectively as the philosophic and the scientific view. The 'first order' philosophic view is the aggregate of human knowledge of the world as known through the senses, interpreted and comprehended by the powers of reason and imagination. The 'second order' scientific view is limited to the material world and the laws that underpin it as revealed by science and its methods. They are both equally *real* – the fact of a snowflake melting in the palm of the hand is every bit as important as the fact of the scientific explanation that its melting involves a loosening of the lattice holding its molecules together. The 'philosophic' view, however, could be said to encompass the scientific, for it not only 'knows' the snowflake melting in the hand as a snowflake, but also the atomic theory of matter and hence its chemical composition.

It would thus seem a mistake to prioritise scientific knowledge as being the more 'real', or to suppose its findings to be the more reliable. But, to put it simply, that is indeed what happened. Before the rise of science, the philosophic view necessarily prevailed, including the religious intimation from contemplating the wonders of the natural world and the richness of the human mind that there was 'something more than can be known'.

From the late eighteenth century onwards the burgeoning success of science would progressively challenge that inference through its ability to 'reduce' the seemingly inscrutable complexities of the natural world to their more readily explicable parts and mechanisms: the earth's secrets surrendered to the geologist's hammer, the intricacies of the fabric of plants and animals to the microscopist's scrutiny, the mysteries of nutrition and metabolism to the analytical techniques of the chemist. Meanwhile, the discovery of the table of chemical elements, the kinetic theory of heat, magnetism and electricity all vastly extended the explanatory powers of science. And, most significant of all, the theory of biological evolution offered a persuasive scientific explanation for that greatest of wonders – the origins and infinite diversity of form and attributes of living things.

The confidence generated by this remorseless expansion in scientific knowledge fostered the belief in its intrinsic superiority over the philosophic view, with the expectation that the universe and everything within it would ultimately be explicable in terms of its material properties alone. Science would become the 'only begetter of truth', its forms of knowledge not only more reliable but more valuable than those of the humanities. This assertion of the priority of the scientific view, known as scientific materialism (or just 'materialism'), marked a watershed in Western civilisation, signalling the way to a future of scientific progress and technical advance while relegating to the past that now superseded philosophical inference of the preceding two thousand years of there being 'more than we can know'. That future, the scientific programme of the twentieth century, would be marked by a progressively ever deeper scientific penetration into the properties of matter, encompassing the two extremes of scale from the vastness of the cosmos to the microscopic cell from which all living things are made. It began to seem as if there might be no limits to its explanatory power.

The genome projects and the Decade of the Brain represent the logical conclusion of that supposition. First, the genome projects were predicated on the assumption that unravelling the Double Helix would reveal 'the secret of life', *as if* a string of chemicals could possibly account for the vast sweep of qualities of the wonders of the living world; and second, the assumption of the Decade of the Brain that those brain scanning techniques would explain the mind, *as if* there could be any equivalence between the electrical firing of neurons and the limitless richness of the internal landscape of human memory, thought and action. In retrospect, both were no more likely to have fulfilled the promise held out for them than to suppose the 'second order' chemical composition of water might account for its diverse 'first order' states of rain, snow, oceans, lakes, rivers and streams as we know them to be.

This necessarily focuses our attention on what that potent 'missing force' must be that might bridge the gap between those two 'orders of reality', with the capacity to conjure the richness of human experience from the bare bones of our genes and brains. This is an even more formidable question than it might appear to be, for along the way those genome projects have also, inadvertently, undermined the credibility of the fundamental premise of what we *do* know about ourselves – that the living world and our uniquely human characteristics are the consequence of a known, scientifically proven, process of biological evolution. Certainly, the defining feature of the history of the universe, as outlined earlier, is of the progressive, creative, evolutionary transformation from the simplest elements of matter to ever higher levels of complexity and organisation. Over aeons of time the clouds of gas in intergalactic space evolved into solar systems such as our own. Subsequently the inhospitable landscape of our earth evolved again into its current life-sustaining biosphere, and so on. Thus the whole history of the cosmos is an evolutionary history. That is indisputable, but the biological theory of evolution goes further, with the claim to know the mechanisms by which the near-infinite diversity of forms of life (including ourselves) might have evolved by a process of random genetic changes from a single common ancestor.

It is, of course, possible that the living world and ourselves did so evolve, and indeed it is difficult to conceive of them not having done so. But the most significant consequence of the findings of the genome projects and neuroscience is the transformation of that foundational evolutionary doctrine into a riddle. The dramatic discovery of Lucy's near-complete skeleton, already described, provides compelling evidence for man's progressive evolutionary ascent over the past five million years. Why then, one might reasonably ask, is there not the slightest hint in the Human Genome of those unique attributes of the upright stance and massively expanded brain that so distinguish us from our primate cousins?

The ramifications of the seemingly disappointing outcomes of the New Genetics and the Decade of the Brain are clearly prodigious, suggesting that we are on the brink of some tectonic shift in our understanding of ourselves. These issues are nowhere more sharply delineated than in an examination of the achievements of the first human civilisation which marked the arrival of our species, *Homo sapiens*, thirty-five thousand years ago.

2

The Ascent of Man: A Riddle in Two Parts

'Alone in that vastness, lit by the feeble beam of our lamps, we were seized by a strange feeling. Everything was so beautiful, so fresh, almost too much so. Time was abolished, as if the tens of thousands of years that separated us from the producers of these paintings no longer existed. It seemed as if they had just created these masterpieces. Suddenly we felt like intruders. Deeply impressed, we were weighed down by the feeling that we were not alone; the artists' souls and spirits surrounded us. We thought we could feel their presence.'

Jean-Marie Chauvet on discovering the world's oldest paintings, from 30,000 BC

The beginning for ourselves, *Homo sapiens* – modern, thoughtful, argumentative, reflective, creative man – can be pinpointed with remarkable accuracy to 35,000 BC, or thereabouts, in south-west Europe. Here, in the shadow of the snow-topped Pyrenees that separate what is now southern France from northern Spain, flourished the first and most enduring of all human civilisations, a vibrant, unified, coherent culture, transmitted from generation to generation for an astonishing twenty-five thousand years. This palaeolithic (Stone Age) civilisation, created by the first truly modern Europeans, was more long-lasting than any that have succeeded it: ten times longer than the 2,500-year reign of the pharaohs in Egypt, twenty-five times longer than the thousand years of Graeco-Roman antiquity.

The historical lineage of our species stretches much further back, into the almost unimaginably distant past of five or six million years ago, but those more ancient predecessors left nothing behind other than some precious and much-argued-over skulls, bones and teeth, and the stone implements, scrapers, blades and axe-heads with which they hunted and butchered their prey.

Homo sapiens, or 'Cromagnon man', as this first representative of our species is known (so named after the Cro-Magnon – 'Big Hole' – rock shelter where his remains were first unearthed in 1868), was something else. His arrival in south-west France signalled a cultural explosion of technological innovation and artistic expression that has characterised the human race ever since. And he was the first, too, to leave behind an image of himself, so though thirty-five thousand years separate us, we can readily make his acquaintance. Fly, or catch the train, to Paris, and take the Metro westwards to the suburban station of St Germain-en-Laye. Emerging from the entrance, you cannot miss the impressive moated château of the Musée Nationale d'Antiquités, home to the largest of all archaeological collections. Few tourists make it this far out of the city, and you may be virtually alone as you stride past the first few display cabinets with their serried ranks of those familiar – if not exactly thrilling – stone implements. And then suddenly, without warning, your eyes are caught by the face of a teenage girl fashioned from the glistening ivory of a mammoth's tusk, so small and delicate she could easily nestle in the palm of your hand (*see overleaf*). Her triangular-shaped face with its long, straight nose and deep-set eyes emerges from a slender, graceful neck framed by flowing locks of braided hair. She is the 'Dame de Brassempouy', the first human portrait, unearthed in 1895 by the French archaeologist Édouard Piette from amongst a pile of mammoth and rhinoceros bones that covered the floor of a cave a few kilometres outside the village of Brassempouy in southern France, after which she is named. Her air of youthful innocence is complemented by a second sculpted object in the same cabinet, of similar size and from the same site, that exemplifies that other timeless image of womanhood – the mature and childbearing. It may only be a broken fragment, but her prominent breasts and fleshy thighs are unmistakably those of a fertile woman.

This youthful teenager and this mature woman, the first images of modern humanity, are both visual and tactile, their polished surfaces testimony to the countless generations whose hands caressed that braided hair and felt those fleshy contours. They subvert the customary perception of man's trajectory from a primitive past to a civilised present by compelling us to recognise how little has changed. The

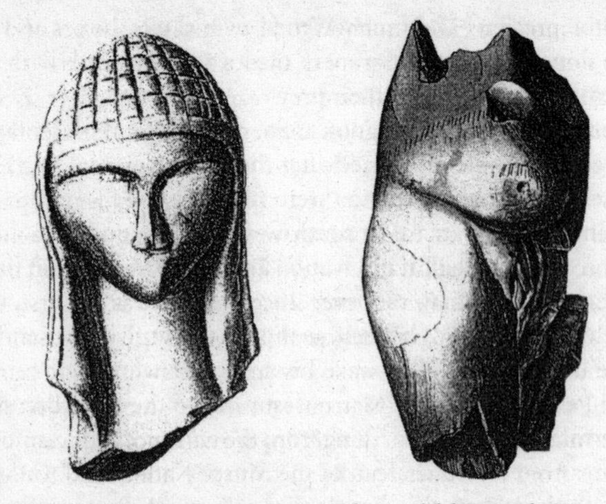

Fig 2-1: *The very earliest human images. The youthful Dame de Brassempouy with her long neck and braided hair, and the sensuous curves of the torso of a mature woman, both unearthed from the same cave in southern France and dated to around 30,000* BC.

cultural history of our species may stretch back thirty-five thousand years, but from its earliest beginnings to the present day it is clearly 'of a piece'.

And there is yet more to the Dame de Brassempouy than this invaluable perspective. Her immediate predecessors in Europe, the beetle-browed, thick-necked Neanderthals (so named after the Neander valley in Germany where their remains were first unearthed), were no more capable of creating so exquisite an object than were the very earliest humans who traversed the savannah plains of Africa several million years previously. Now, those Neanderthals had many virtues. They were tough and intelligent enough to survive for a quarter of a million years in the hostile environments of the recurring Ice Ages that periodically swept across the continent, and they had a brain capacity slightly larger than our own. But they left not a single such image behind. The Dame de Brassempouy thus focuses our attention with exceptional clarity on that most important of questions: What happened in the transition to modern man? What is it that sets us apart, why should we be so different?

The Cromagnons' arrival in south-western Europe was the culmination of an unexplained diaspora that 100,000 years earlier had impelled modern *Homo sapiens* to leave his African homeland and spread outwards to every corner of the earth. It was cold, of course, as throughout the tens of thousands of years of Cromagnon civilisation the ice cap several hundred miles to the north expanded and retreated. But they found shelter from the icy winds in the rocky south-facing valleys of the Dordogne and the Pyrenees. They had fire to warm themselves and animal furs for clothing, sewn together with the aid of ivory needles and held in place by exquisite ivory buttons. They lived in communities of several hundred spread out in separate dwelling places, and with a total population of probably little more than twenty thousand. They danced, as we know from the swinging breasts of an exquisite thirty-thousand-year-old statuette of a naked woman, and played music, fashioning drums from mammoth bones, clicking castanets from jawbones and flutes from the hollow bones of birds, which, with a whistle head attached, can be made to produce strong, clear notes. They wore jewellery and beads made from a few highly prized materials – certain types of seashells and animal teeth – which they traded over large distances. And they were great technical innovators. While their predecessors' stone tools had scarcely changed in a million years, the Cromagnons prodigiously extended their sources of food supply by inventing both the spear-thrower and the harpoon. They invented oil lamps to illuminate the interiors of their caves, the drill that could put an 'eye' in a needle, and rope to bind their tents together.

And they had a passion for art. 'We are justified in asserting they devoted themselves, intensely and continuously, to the creation of pictorial, graphic and sculptural works,' writes the Italian art historian Paolo Graziosi. This is not the conventional version of primitive Stone Age art, where 'stick' men pursue their quarry with bows and spears, but is comparable to the art of the Italian Renaissance, with a naturalistic style that 'sought to express reality in its deep unchanging essence'. Their powers of observation were so acute that 'we know, for example, that the extinct rhinoceros of Ice Age Europe was adorned with a shaggy coat', writes Ian Tattersall of the American Museum of Natural History, and that the extraordinary *Megalocerus giganticus*, a deer with vast antlers, had a darkly coloured hump behind its shoulders.

The Cromagnons' artistic legacy takes two forms: 'portable' art, mostly sculptures and engravings on ivory and antler horn; and the distinctly 'non-portable' vast frescoes that covered the walls and ceilings of their cavernous cathedrals concealed in the depths of the mountainsides, in which they 'mastered the problems of presenting three dimensions in two, and in giving a sensation of movement'.

And what movement! As archaeologist John Pfeiffer recalls on first glimpsing the 'incomparable splendour' of the painted caves at Lascaux in southern France:

> It is pitch dark inside, and then the lights are turned on. Without prelude, before the eye has had a chance to look at any single feature, you see it whole, painted in red, black and yellow, a burst of animals, a procession dominated by huge creatures with horns. The animals form two lines converging from left and right, seeming to stream into a funnel mouth, towards and into a dark hole which marks the way into a deeper gallery.

It is not possible to convey the full range of the Cromagnons' artistic virtuosity, so three striking examples must suffice. The first is the sculpted handle of a spear-thrower, fashioned from a reindeer's antler, that shows a young ibex looking round at a large faecal stool emerging from its rectum, on which two birds are perched. The tautness of the animal's neck muscles is beautifully conveyed in this humorous image, which must have been popular as several others, virtually identical, have since been discovered.

Next comes a fresco painting of a pride of lions from the Chauvet cave of the opening quotation to this chapter, whose 'richly embellished chambers' also feature mammoth, rhinoceros and an 'exquisitely painted'

Fig 2-2: *The Cromagnons were great technical innovators – the lamp, the drill, fish-hook, harpoon and spear-thrower – which, with their 'passion for art', they embellished with beautiful naturalistic engravings.*

Fig 2-3: *The Cromagnons' sophisticated artistic style is apparent in this fresco of a pride of lions from the Chauvet cave in southern France, where the heavy paint-strokes beneath the neck convey a sense of perspective.*

panel of horses' heads. But these lions are the most impressive of all, showing how the Cromagnons had mastered the three-dimensional sense of perspective, with heavy paint-strokes beneath the neck adding depth to the image.

Thirdly there is a bison's head (*see overleaf*), full of gravitas, sculpted from clay, from around 15,000 BC, that would be a masterpiece in any age. One can hardly imagine that it was created 14,500 years before the Golden Age of sculpture of classical Athens. Indeed, when compared to the sculpted head of an ox being led in sacrificial procession on the Parthenon frieze, one could almost be forgiven for thinking they were rendered by the same hand.

So there we have it, three artistic masterpieces, each of which conveys something of the profundity of the mind of these first representatives of our species, their humour and pathos, and their deep appreciation of the character of the animals with which they shared the world and which they so masterfully portrayed.

Fig 2-4: *Compare and contrast: the bison's head (top) sculpted in a cave in southern France in 15,000* BC, *and the ox (bottom) from the Parthenon frieze of 430* BC.

For the archaeologists of the nineteenth century, few things were quite as perplexing as the possibility that these wonderful expressions of human intelligence could have been created by Stone Age cave dwellers. It seemed inconceivable that man could have been capable of such artistic virtuosity in so distant a past, so when the Marquis de Sautola stumbled across the first of the painted caves at Altamira in northern Spain

– after his eight-year-old daughter drew his attention to a parade of bison on the ceiling with the famous phrase '*Mira, Papa, bueyez!*' (Look, Papa, oxen!') – no one believed him. His lecture to the International Congress of Archaeology in Lisbon in 1880 in which he described his findings was 'met with incredulity and an abrupt and contemptuous dismissal'. The Altamira paintings were never acknowledged as authentic in his lifetime – rather, their exceptional quality was presumptive evidence that he must have faked them. The Marquis died a disillusioned man, yet any condemnation of his harsh treatment by his archaeological contemporaries is a judgement of hindsight. Their scepticism was not unreasonable, given that the proposition that Stone Age man might have been capable of creating such great art itself seemed unreasonable.

Nowadays we know better. The sensational recent discoveries of man's earliest ancestors – in particular the two near-complete fossilised skeletons, 'Lucy' and 'Turkana Boy' – mark the first two distinct stages of Man's Ascent, his decision to *stand upright* and his *prodigiously enlarging brain*. The 'cultural explosion' of Cromagnon man's artistic achievement marks the culminating phase of that evolutionary trajectory, determined by the third of those distinctly human attributes – the *faculty of language*. And yet, the drama of that evolutionary trajectory now appears, in the light of the findings of the New Genetics and the Decade of the Brain, more perplexing even than it would have seemed to those sceptical nineteenth-century archaeologists. This is 'the riddle of the ascent of man'.

The common understanding of man's evolutionary heritage begins with Charles Darwin's *On the Origin of Species* of 1859, extended to incorporate 'ourselves' in *The Descent of Man*, published twelve years later, with its central claim that the near-infinite diversity of shape, form and attributes of living things all evolved from the first and simplest form of life, self-assembled from 'all sorts of ammonia and phosphoric salts' in 'some warm little pond' on the earth's surface several billion years ago. The modern interpretation of Darwin's theory is, briefly, as follows. The major determinants of what makes a fish a fish, or a bird a bird, are the instructions carried within the twenty thousand (plus or minus) genes formed by the sequence of just four chemicals (best imagined, as suggested, as four coloured discs, green, red, blue and yellow) strung out along the Double Helix within the nucleus of each and every cell. These genetic instructions are then passed on in the sperm and egg at the moment of conception to ensure that the offspring of fish will be fish and birds, birds. Those individual genes replicate themselves with astonishing accuracy every time the cell divides, but very occasionally a mistake, or 'mutation', may creep in: so a green disc (say) is substituted for a red one, thus subtly altering the genetic instructions. Most of the time this does not matter, or is detrimental, but very occasionally the 'chance mutation' in those genetic instructions may confer some biological advantage, maximising the carriers' chances of survival in the struggle for existence. Their offspring in turn are likely to inherit their parent's advantageous genetic variation, and as the process continues from generation to generation, the characteristics of species will be gradually transformed, step by step, in favour of those which are best suited (or 'adapted') to their environment. Thus fish are adapted to life underwater because over millions of generations 'nature' has 'selected' (hence 'natural selection') those whose random changes in their genes have maximised their swimming potential, while birds are good at flying because the same process has maximised their aerodynamic capabilities. Put another way, all of life has 'descended with modification' from that common ancestor.

And man, Darwin argued in *The Descent of Man*, is no exception. Indeed, there is probably no more persuasive evidence for his evolutionary theory than the striking physical similarities between man and his primate cousins, which point inexorably to their having 'descended by modification' from some common ape-like ancestor. Man has survived and prospered because nature, in 'selecting' the genetic mutations that would cause him to stand upright and acquire that much larger brain, conferred so considerable a biological advantage as to maximise his chances of survival. Certainly man's much superior intellectual faculties might seem to set him apart. Nonetheless, our primate cousins, like ourselves, exhibit similar emotions of jealousy, suspicion and gratitude; they make choices, recall past events and are capable (to a degree) of reason. Hence the superiority of the human mind, Darwin argued, represents a continuum – it is a difference of 'degree but not of kind'.

Details aside, one single, powerful image captures this profoundly influential interpretation of man's origins. Darwin's close friend and advocate Thomas Huxley, in an illustration to his book *Evidence as to Man's Place in Nature* (1863), placed the skeletons of chimpanzees, gorillas and man in sequence, transforming their striking physical similarities into a powerful narrative of the rise of man from knuckle-walking chimp to upstanding *Homo sapiens*. The same image, expanded with a series of 'hominid' intermediaries, and reproduced (often humorously) in numerous different guises, would become one of the most familiar, and certainly influential, icons of the twentieth century. The 'Descent' of man by modification from his ape-like ancestors, it implied, was in reality the story of his 'Ascent' to his preeminent position in the grand order of life. And so the major archaeological discoveries of the last fifty years have shown it to be.

Fig 2-5: *The Ascent of Man. Thomas Huxley's series of skeletons of gibbon, orangutan, chimpanzee, gorilla and human. The same inference of a gradualist, progressive evolutionary transformation from ape to man would subsequently be captured in numerous popular illustrations and cartoons.*

The discovery of the fossilised bones of Cromagnon man in 1868, together with those of his beetle-browed Neanderthal predecessors, would roll back man's evolutionary history 200,000 years or more. It was not however till the 1930s that the first evidence of the more distant stages would begin to emerge, and not till the 1970s, when the first of two near-complete fossilised skeletons was unearthed in the harsh landscape of central Africa, that Darwin's hypothesised transition from that ape-like common ancestor to *Homo sapiens* would be vindicated.

Those two near-complete skeletons were, first, the three-and-a-half-million-year-old 'Lucy', or *Australopithecus afarensis* to give her her scientific name, whose discovery had such a powerful and emotional effect on Donald Johanson and Tom Gray ('...we hugged each other, sweaty and smelly, howling and hugging in the heat-shimmering gravel'), as already described. One set of fossilised remains can look remarkably like another, but the vital clue, and the source of their exhilaration, lies in the sharp upward angle of the head of the femur, or upper thigh bone, that locates the centre of gravity of the human skeleton in the small area enclosed by two feet placed together – confirming that Lucy was the first of our most distant ancestors to stand upright.

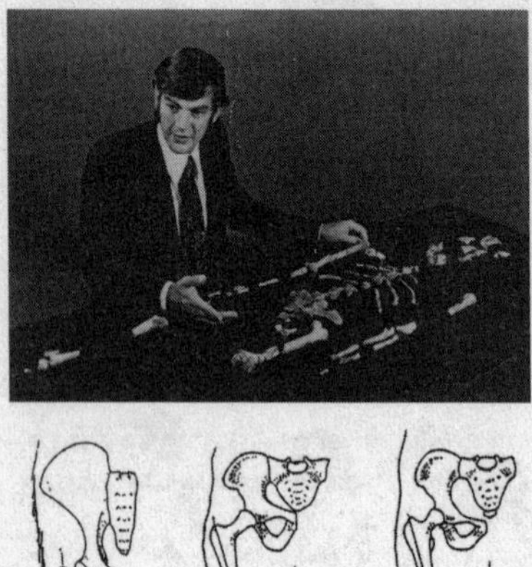

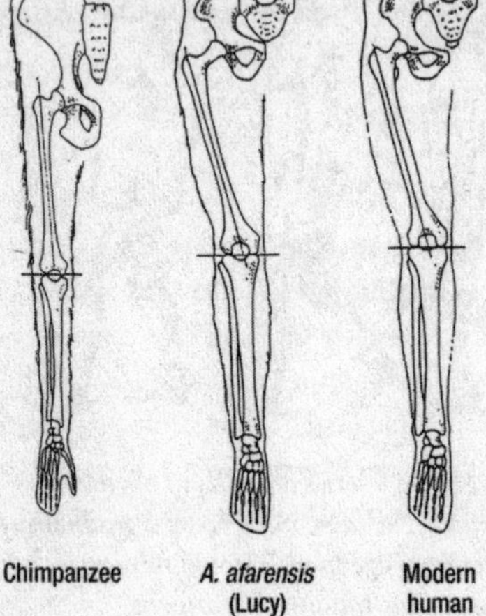

Fig 2-6: *Donald Johanson demonstrates the distinctive feature of Lucy's skeleton. The two vital clues to her human status lie in the sharp upward angle of the hip joint and her distinctively shaped pelvis, that together relocate the body's centre of gravity over the feet – thus allowing her to stand upright.*

Lucy's novel method of locomotion would be confirmed soon after with the discovery of an amazing series of three sets of footprints left behind in the volcanic ash of the Laetoli region of northern Tanzania three and a half million years ago, that provide a most moving insight into the human relationships of those distant ancestors.

Fig 2-7: *The Laetoli footprints of an adult and two children making tracks across ground covered by volcanic ash. The prints exhibit the same distribution of forces as in modern humans, with the main stress on the ball of the foot and the heel.*

'The footprint records a normal positioning of the left and right feet with human-like big toes,' writes British neurologist John Eccles.

> The special feature is that one individual followed the other, placing its feet accurately in the preceding footsteps. The third was smaller and walked closely to the left following the slightly wavy walk of the larger. We can interpret this as showing that two of the group [mother and child?] walked together holding hands, while another [sibling?] followed accurately placing its feet into the footsteps of the leader. We are given a privileged view of a family taking a walk on the newly formed volcanic ash 3.6 million years ago, just as we might do on soft sand left by the receding tide!

Ten years later, in 1984, near the ancient Lake Turkana in Kenya, the famous palaeontologist Richard Leakey discovered a further near-complete skeleton, 'Turkana Boy', an adolescent of the species *Homo erectus* from around 1.6 million years ago, with a skull intermediate in size between Lucy and ourselves – reflecting that second unique evolutionary characteristic of the human species, his prodigiously enlarging brain.

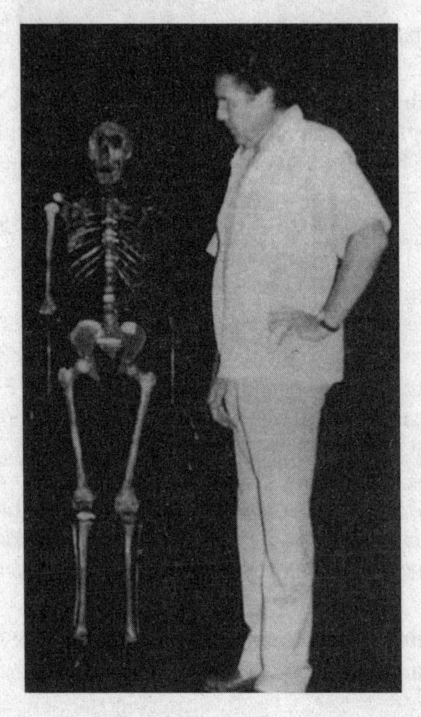

Fig 2-8: *The director of excavations, Alan Walker, stands beside the reconstructed skeleton of the adolescent Turkana Boy, whose skull size, intermittent between Lucy and ourselves, reflects the progressive evolutionary expansion of the brain.*

Turkana Boy's 'people', *Homo erectus*, were the first to make tools, reflecting that manual dexterity which is a further unique attribute of humans, made possible by the seemingly trivial evolutionary advantage of lengthening the thumb by an extra inch to make it 'opposable', allowing it to 'speak to' the other four digits.

> 'In man the most precise function that the hand is capable of is to place the tip of the thumb in opposition to the tip of the index finger, so they make maximum contact,' writes the British anatomist John Napier. 'In this position small objects can be manipulated with an unlimited potential for fine pressure adjustments or minute directional corrections. This is the hallmark of mankind. No non-human primate can replicate it.

This extra inch of the human thumb transforms the *grasping* power of the hand of our primate cousins into the vast repertoire of the *gripping* precision of the human hand that would, eventually, allow man to paint and sculpt and record, through the written word, his experiences, without which his history would have disappeared completely into the dark abyss of time.

We cannot know precisely how or why the enlarging brain of *Homo erectus* released the hand's (till then, hidden) potential of both *grasping* and *gripping*, but we can see its consequences readily enough in his stone tools. Palaeontologists who have taught themselves the technique of stone napping (as it is known) discovered that the necessary skill lies in finding the right-shaped 'core' stone, which is then percussed by a hammer stone at precisely the right angle and with a controlled degree of force so as to produce fragments of the right size and sharp enough to cut open the skin of his prey.

There is something immensely moving about the diminutive Lucy, no more than five feet tall, and the strapping Turkana Boy. Their bones may be silent, but nonetheless they speak to us across those aeons of time. What, one wonders, when they looked upwards at that clear African sky at night, did they make of its thousands of shimmering stars and the drama of the waxing and waning of the moon?

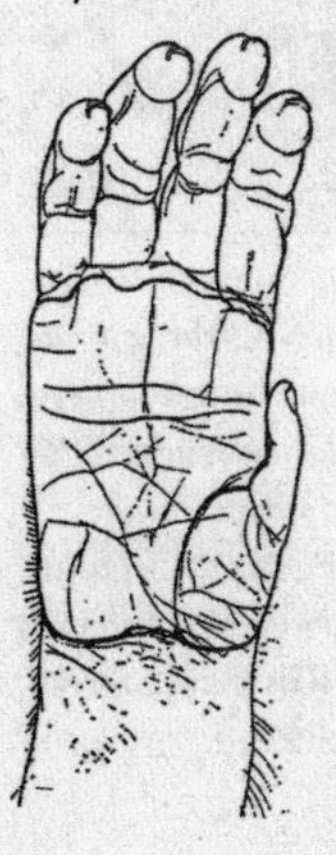

Fig 2-9: *The hallmark of mankind. The extra inch of the human thumb complements the grasping power of the primate hand with the breathtaking precision necessary to thread a needle or play the flute.*

It is impossible to exaggerate the importance of those skeletal remains to our understanding of evolutionary heritage, confirming the linear sequence of our predecessors just as Darwin had postulated. Five million years ago the antecedents of Lucy and her tribe forsook the safety of the forest to walk upright on the savannah of central Africa. Two million years passed, and the ever-expanding brain of *Homo erectus* allowed for the incremental increase in intelligence necessary to fashion tools from stone, and to undertake those extraordinary migrations of tens of thousands of miles that would take him through what is now the Middle East into Asia, and as far as northern China and Indonesia. Then a further million and a half years elapsed before the emergence of *Homo sapiens*, who from his base in Africa would undertake a second great wave of global migration, this time as far as Australia, across the Bering Straits to America, and up into Europe and the southern Pyrenees, where Cromagnon man would found the first human civilisation.

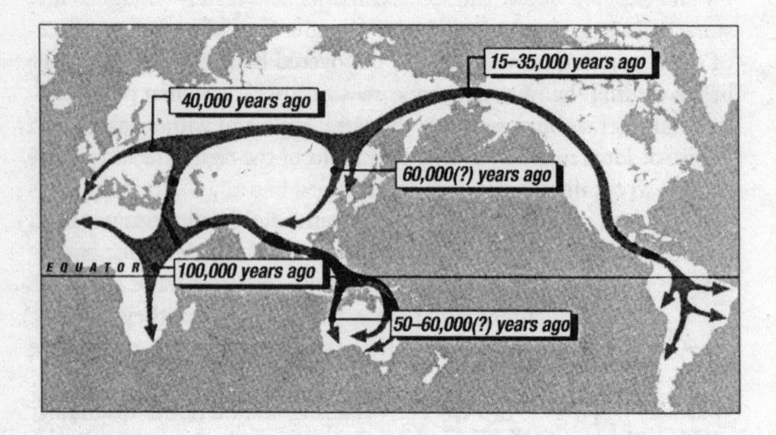

Fig 2-10: *Man's far-flung diaspora. During the last and most far-reaching of the successive waves of human migration out of Africa, modern man,* Homo sapiens, *colonised the globe, displacing the preceding indigenous populations such as the Neanderthals in Europe. Along the way they would have to surmount the insuperable geographical obstacles of hundreds of miles of open sea that separate Indonesia from the Australian continent, and the freezing wastes of the temporary land bridge linking Asia with the Americas created by the fall in the sea level caused by the Ice Age that lasted from 30,000 to 20,000* BC.

Thanks to Donald Johanson, Richard Leakey and many others, we now possess the factual evidence for man's evolutionary ascent, culminating in those images of deer and bison on the ceiling of that Altamira cave whose technical virtuosity so astonished the Marquis de Sautola. There is no deep mystery about our origins. The massively overwhelming prevailing view taught in virtually every school and university in the Western world insists that Darwin's evolutionary theory of natural selection explains us and our ancestors. 'Our own existence [that] once presented the greatest of all mysteries is a mystery no longer. Darwin solved it,' observes the evolutionary biologist Richard Dawkins. 'We no longer have to resort to superstition . . . it is absolutely safe to say that anyone who claims not to believe in evolution is ignorant, stupid or insane.' And how could it not be so? What other conceivable explanation might there be? There is none. That, one would think, should be the end of the matter.

And yet, the more that time has passed since those definitive discoveries of Lucy and Turkana Boy, the more perplexing that evolutionary trajectory seems to be. The Ascent of Man captured in Thomas Huxley's famous image is no longer a theoretical idea: it is concretely realised in Lucy's sharp-angled femur and Turkana Boy's much larger skull; but the more one reflects on what is involved in standing upright or acquiring a larger brain, the less convincing Darwin's proposed mechanism of natural selection appears to be. Further, the suddenness of the cultural explosion that signalled the arrival of Cromagnon man argues against a progressive, gradualist evolutionary transformation. It suggests rather some dramatic event – as if a switch were thrown, the curtain rose, and there was man at the centre of the stage of world events. The findings of the New Genetics and the Decade of the Brain make it much more difficult to set such doubts aside. The trivial *genetic* differences that separate our primate cousins from ourselves seem quite insufficient to account for those *physical* differences that set us apart. Similarly, the elusive workings of the human brain would seem to defy any simple evolutionary explanation.

THE RIDDLE OF 'THE ASCENT'

Part 1: Setting Out

There are at least half a dozen speculative evolutionary reasons for why Lucy and her kind might have wished to stand upright, and the advantage in doing so: the better to see potential predators, to carry her dependent offspring, or, as Darwin himself supposed, 'standing on two legs would permit an ape-like predecessor to brace itself by holding on to a branch with one arm as he grabbed the fruit with another'. But the most schematic anatomical comparison with our primate cousins reveals the prodigious difficulties of this novel form of locomotion. The knuckle-walking chimp has four powerful, pillar-like limbs, providing a large rectangular basis of support, with the centre of gravity solidly located in the middle of the torso. For Lucy, the centre of gravity shifts to a small area enclosed by her two feet, and with the bulk of her weight (her head and torso) in the upper part of the body, exacerbating her tendency to topple over. While the chimpanzee might be compared to a solid, four-legged table, Lucy's upright frame, like an unsupported pole balancing a heavy ball (her head), must constantly defy the laws of gravity.

So how did she come to stand upright? The main anatomical transformations pivot around the pelvis, where the powerful *gluteus maximus* of our buttocks, a minor player in our primate cousins, pull the human form upright like a drawbridge. This novel stance must then be held in place by a redesign of several other muscles that fulfil a *propulsive* function in our primate cousins, but need to act as *stabilisers* of the human skeleton. For that to happen, the bony pelvis to which they are all attached must itself undergo a major redesign, being first pulled up and back, then shortened and widened.

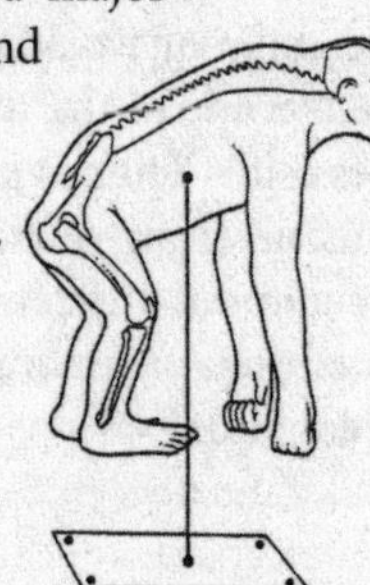

Fig 2-11: *The hazards of the upright stance. The pole-like upright stance of Lucy and her kind must, when compared to the four-square solidity of the chimpanzee, constantly defy the laws of gravity.*

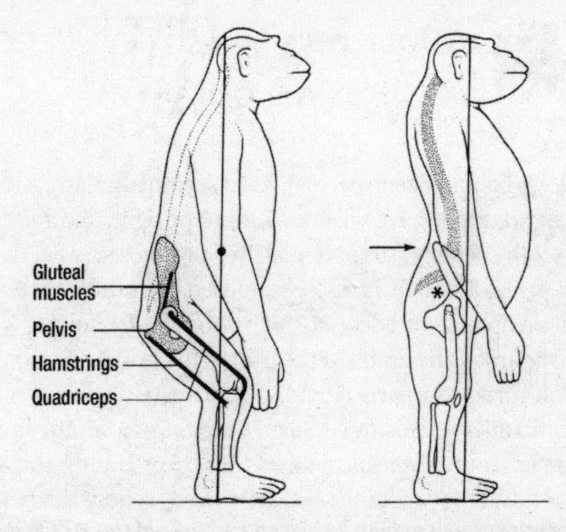

Fig 2-12: *The several anatomical transformations necessary for the upright stance include a radically redesigned pelvis, a major increase in the bulk of the gluteal (buttock) muscles and the repositioning of their attachments to pull the skeletal frame upwards, and a concave curvature of the lower lumbar spine.*

This redesign of the pelvis and its stabilising muscles entails a further series of knock-on effects: the *skull* must now be repositioned directly over the erect human frame, while the *vertebrae* of the spinal column must become progressively wider as they descend, to sustain the weight pressing upon them. The *head of the femur* (as noted) must be angled inwards, the *ligaments of the knee* strengthened to 'lock' into position, while the *foot*, particularly the big toe, must undergo a dozen anatomical changes to provide a firm basis of support. The *arms*, that no longer need to swing through branches, become shortened, while the *legs*, now proportionally too short relative to the body, must be lengthened – but by how much? There is, it would seem, an 'ideal' length that creates a pendulum-type movement of the legs, like the pendulum of a clock, where walking becomes almost automatic, the combination of the force of gravity and inertia carrying the body forward with hardly any intentional muscular effort. 'The human frame is built for walking,' observes the biomechanist Tad McGeer. 'It has both the right kinematics and the right dynamics – so much so in fact that our legs are capable of walking without any motor control.'

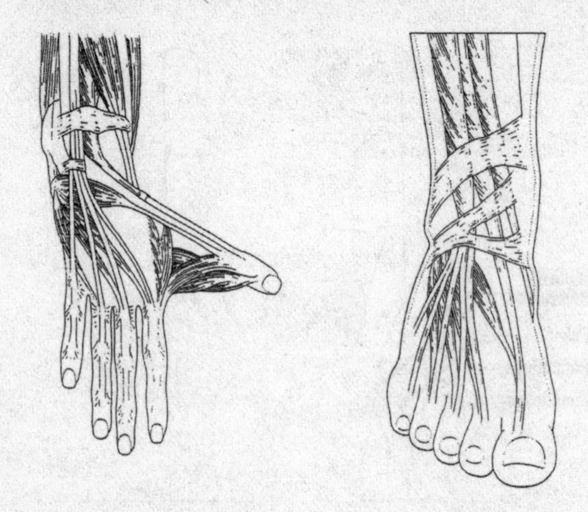

Fig 2-13: *The primate and human foot, despite superficial similarities, reflect their very different purposes – the primate foot grips, whereas the human foot must sustain the downward pressure of the erect human frame (measured in kilotons per square inch when running or walking), with the big toe acting as a lever, making it an 'elastic, mobile, dynamic' organ. Note the contrast in the positioning of the muscles and ligaments that bind them, which is even more marked on the under-surface of the foot.*

And that shortening of the arms and lengthening of the legs for walking gives the symmetry and harmony that reflect the hidden laws of geometric proportion captured by Leonardo da Vinci's famous 'Vitruvian' image of man, his span matching his height, encompassed within the two most elemental of shapes – the circle and the square. 'It is impossible to exaggerate what this simple-looking proposition meant to Renaissance man,' the art historian Kenneth Clark observed. 'It was the foundation of their whole philosophy,' where man was 'the measure of all things'.

These, however, are merely the obvious anatomical changes, as Lucy would still be quite unable to stand upright without both a rewiring of her nervous system to cope with the flood of feedback information from the millions of sensors monitoring the relative position of the bones and muscles, *and* a more sophisticated circulatory system. Thus, those obvious similarities that so impressed Darwin and Huxley conceal a myriad of hidden but necessary modifications, because there

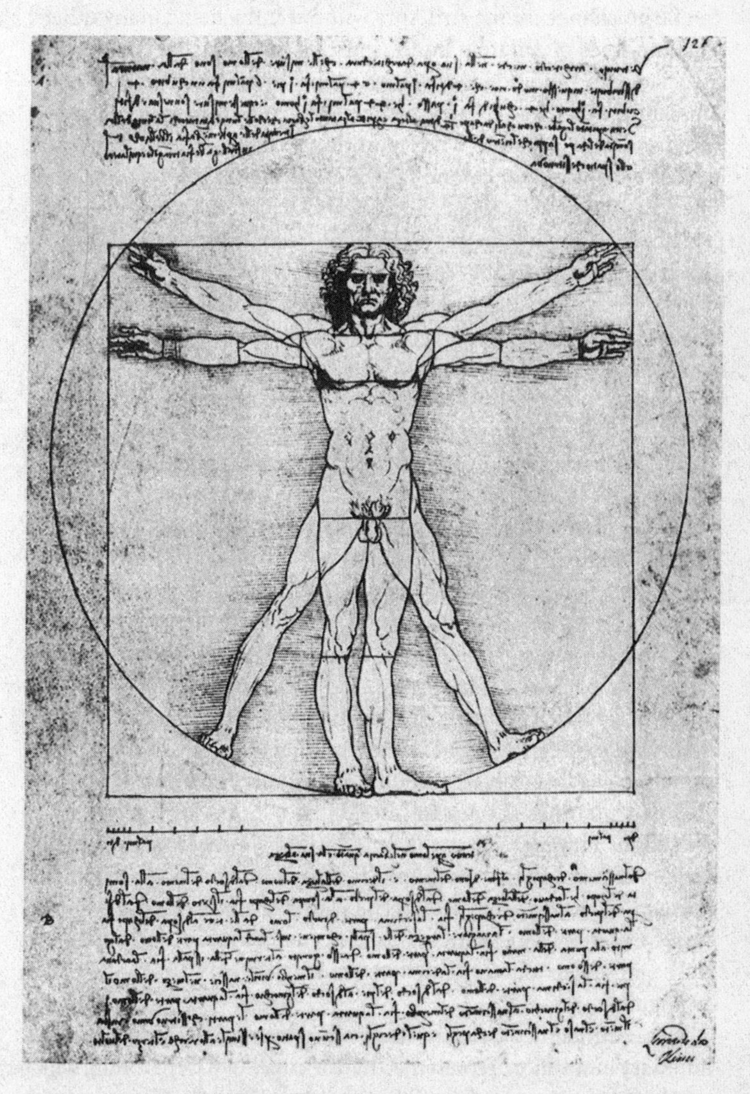

Fig 2-14: *Leonardo's famous 'Vitruvian Man', whose proportions, with arms outstretched, encompass the two elemental shapes of square and circle, 'without which it is impossible to create anything'.*

can be no change in one structure without influencing many others. The hundreds of different bones, muscles and joints are 'inseparably associated and moulded with each other', the distinguished twentieth-century biologist D'Arcy Wentworth Thompson would observe. 'They are only separate entities in this limited sense that they are *part* of a whole – that can no longer exist when it loses its *composite* integrity.'

Lucy was already an experienced 'bipedal walker', as shown by those emotive footprints in the volcanic ash – so we must presuppose, if Darwin's evolutionary theory is correct, numerous preceding species of hominids marking out those anatomical changes from the rock-like stability of the knuckle-walking primates to her own upstanding, pole-like form. We can only speculate how those changes occurred. The strengthening of the gluteus muscle was essential to 'raise the drawbridge' – but it would have been quite unable to do so without the simultaneous redesign of the bones of the pelvis and upper thigh, the ligaments to lock the knees, the adaptation of the foot to standing upright, and so on. Thus the biological advantage of 'freeing the hands' would be more than offset by the profound instability of any transitional species, that without this full house of anatomical changes would have had a stuttering, shuffling gait – vulnerable prey to any hungry carnivore it encountered when tottering across the savannah. Put another way, the necessity for these many anatomical changes confirms what one would expect: the upright stance is staggeringly difficult to pull off, which is presumably why no other species has attempted it. Standing upright is, on reflection, a rather bizarre thing to do, and would seem to require a sudden and dramatic wholescale 'redesign' that is clearly incompatible with Darwin's proposed mechanism of a gradualist evolutionary transformation. Lucy's pivotal role in man's evolutionary ascent as the beginning or anchor of that upward trajectory would seem highly ambiguous.

These difficulties seem less acute when we turn to that second evolutionary innovation, represented by Turkana Boy's larger brain, which would at least have conferred the obvious advantage of greater intelligence – and progressively so, each incremental increase in brain size and intelligence furthering his chances of survival. Except, there is no direct evidence for the benefits of that greater intelligence, other than those stone tools which, for all their technical ingenuity, remained virtually unchanged for two million years. The human brain started to increase in size, and continued to do so over a period of several million years, with precious little to show for it until right at the end, with that extraordinary intellectual leap of the Cromagnons' 'cultural explosion'. Why, one might reasonably ask, should man's evolutionary progress equip him with powers he would not realise for so long?

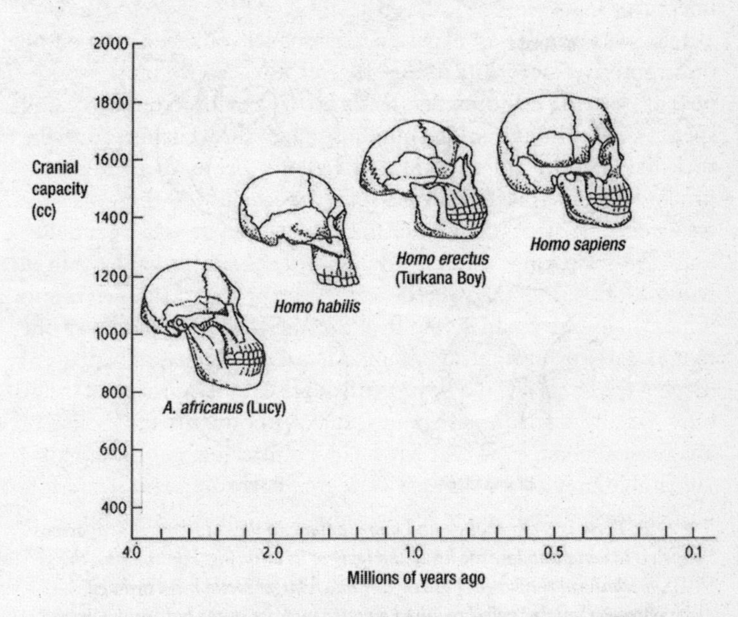

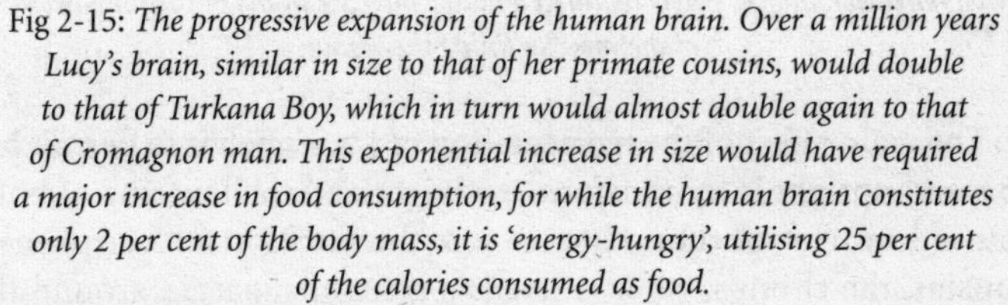

Fig 2-15: *The progressive expansion of the human brain. Over a million years Lucy's brain, similar in size to that of her primate cousins, would double to that of Turkana Boy, which in turn would almost double again to that of Cromagnon man. This exponential increase in size would have required a major increase in food consumption, for while the human brain constitutes only 2 per cent of the body mass, it is 'energy-hungry', utilising 25 per cent of the calories consumed as food.*

This is no mere rhetorical question, for man, during his Ascent, paid a heavy price for that expanding brain, which together with standing upright would massively increase the risk of obstetric catastrophe in childbirth – a fact that could scarcely be more biologically disadvantageous to the survival of his species. A simple diagram of the foetal head passing down through the bony pelvis explains all.

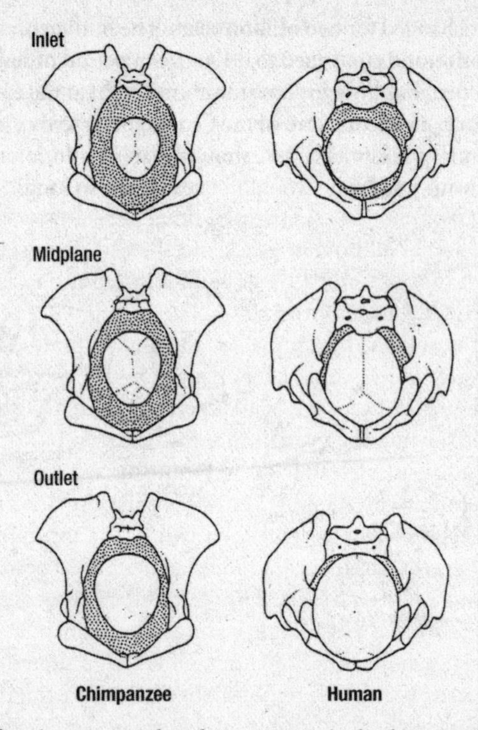

Fig 2-16: *There is room enough and more to spare in the chimpanzee's capacious pelvis to accommodate the head and body of its offspring. For humans, the mechanical challenge of forcing the much larger foetal head through a narrower, angled pelvis required a major increase in the bulk and power of the muscles lining the womb.*

The main effect of the reorientation of Lucy's pelvis to permit her to stand upright is to transform a straight and shallow ring of bone into a deep, curved tube. First we note the situation for our primate cousins, the chimps, where there is a generous margin around the foetal head. Next we see the much tighter fit of Lucy's pelvis, which becomes a potentially lethal crush for both mother and infant. And further, the foetal head must now overcome the greater resistance of the mother's much more powerful pelvic muscles, strengthened to retain her internal organs within the abdomen against the downward force of gravity. A million years on, and the 'bigger brain' of Turkana Boy further compounds these difficulties, so now it requires massive protracted (and very painful) contractions of the muscles of the uterus (ten times more powerful than in other mammals) to force the foetus down that 'deep, curved tube', causing potential damage to the pelvic muscles, bowel and bladder. The predictable consequence of all this is that while the chimpanzee can give birth on her own, almost without breaking stride, humans right from the beginning would have required the assistance of others to support them in this most traumatic of all human experiences – with a mortality rate of 100 per cent for both mother and child in the not unusual circumstance of obstructed labour.

And that is only the beginning, for the human brain continued to increase in size, which would have created an insurmountable obstacle to reproduction were it not for the extraordinary evolutionary 'solution' of slowing the growth rate of the foetal brain within the womb, and accelerating it afterwards. From Turkana Boy onwards, the human newborn, with its now relatively immature brain at birth, is completely helpless, and it will take a further year and a half before it starts to acquire the sort of motor skills that permit the newborn infant chimp to hang on to its mother's back. And that dependency in turn would require that further unique feature of humanity, the long-term pair bond between mother and father, hinted at in those footsteps in the volcanic ash, to share the responsibility for carrying, caring for and feeding their dependent offspring.

It can be difficult to appreciate the devastating implications of these reflections. The Ascent of Man from knuckle-walking chimp to upright human seems so logical and progressive as to be almost self-evident, yet it conceals events that are without precedent in the whole of biology. The only consolation would be that man must have evolved *somehow*, but then the hope of understanding *how* would seem to evaporate with the revelation of the near-equivalence of the human and chimp genomes. There is nothing to suggest the major genetic mutations one would expect to account for the upright stance or that massively enlarged brain – leading the head of the chimp Genome Project to concede, as already cited, somewhat limply: 'Part of the secret is hidden in there, we don't know what it is yet.' Or as a fellow researcher put it, rather more bluntly: 'You could write everything we know about the genetic differences in a one-sentence article.' The reports in 2005 of a family in southern Turkey with a bizarre genetic defect that caused them to walk on all fours suggested, according to Professor Uner Tan of Cukurova University, the breakthrough of 'a live model for human evolution'. Perhaps, but then perhaps not, as the anatomy of the family's bones and muscles was otherwise entirely human, so with relatively short arms and long legs, their ungainly quadrupedal locomotion only served to emphasise the 'full house' of anatomical transformation necessary for the upright stance.

So, while the equivalence of the human and chimp genomes provides the most tantalising evidence for our close relatedness, it offers not the slightest hint of how that evolutionary transformation came about – but rather appears to cut us off from our immediate antecedents entirely. The archaeological discoveries of the last fifty years have, along with Lucy and Turkana Boy, identified an estimated twenty or more antecedent species, and while it is obviously tempting to place them in a linear sequence, where Lucy begat Turkana Boy begat Neanderthal man begat *Homo sapiens*, that scenario no longer holds. Instead we are left with a bush of many branches – without there being a central trunk linking them all together.

> 'Over the past five million years new hominid species have regularly emerged, competed, co-existed, colonised the environment and succeeded or failed,' writes palaeontologist Ian Tattersall. 'We have only the dimmest of perceptions of how this dramatic history of innovation and interaction unfolded, but it is evident that our species is simply one more of its many terminal twigs.'

The methods of the New Genetics have confirmed that all the human races – Negroes, Caucasians, Asians and so on – are genetically identical, thus all descendants of an entirely novel species, *Homo sapiens*, ourselves, who emerged it is presumed in east or south Africa in 120,000 BC before spreading out to colonise the world. But that leaves the 'terminal twig' of ourselves suspended in limbo, with no obvious attachment to those earlier branches of that evolutionary bush. The account of ourselves which until recently seemed so clear now seems permeated with a sense of the deeply inexplicable – whose implications we will return to after considering the second aspect of the riddle of that evolutionary trajectory: the Cromagnons with their 'passion for art'.

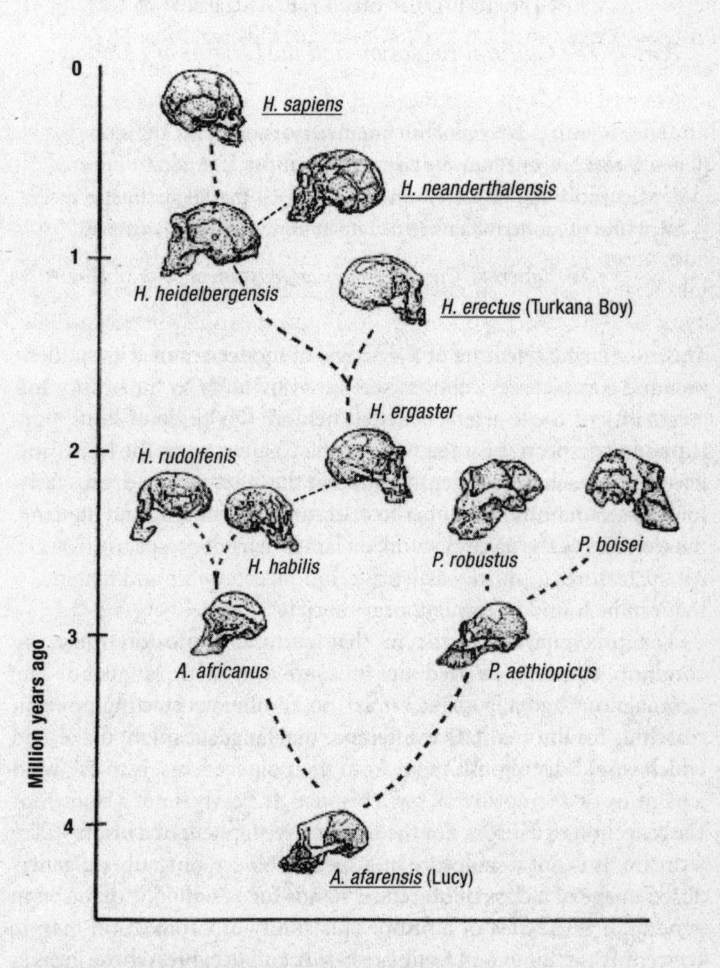

Fig 2-17: *The riddle of human ancestry. The techniques of the New Genetics have confirmed that the diverse races of* Homo sapiens *– Caucasian, Negroid and Asiatic – are all descended from the same common stock, genetically quite distinct from their chronologically most immediate antecedents, the Neanderthals. The enigma of man's relationship to his predecessors is illustrated here by the dotted line, the conventional depiction of uncertain provenance in family trees.*

THE RIDDLE OF THE ASCENT

Part 2: The Cultural Explosion and the Origins of Language

> '*Homo sapiens* is not simply an improved version of his ancestors, but a new concept, qualitatively distinct from them . . . A totally unprecedented entity has appeared on the earth. All the major intellectual attributes of modern man are tied up in some way with language.'
>
> *Ian Tattersall, Curator, American Museum of Natural History*

The most striking feature of the arrival of modern man is its suddenness and completeness, epitomised most obviously by the beauty and originality of those artefacts he left behind: the 'pride of lions' portrayed in perspective on the walls of the Chauvet cave; the beads and jewellery for self-adornment in this and the 'next' world; drums fashioned from mammoths' bones to celebrate, with singing and dancing, the wonders of the natural world; oil lamps, harpoons, spear-throwers. All the features in short – artistic, technical, economic and religious – that can be found in contemporary society.

The precipitating factor in that cultural explosion must, by common consent, be tied up in some way with language. The Cromagnons had a 'passion for art', so an obvious starting point in searching for the qualitative difference that language might make, and which would distinguish them from their antecedents, is to ask what a painting or a sculpture of, say, a bison, *is*. It clearly is not a bison, nor the reflection of a bison, nor the imaginative figment of a bison – as in a dream. It is not a sculpture of a specific object, but rather a generalised image of a class of objects: it stands for, is symbolic of, bison in general. It is the *idea* of a bison. This ability of Cromagnon man to conceptualise things and feelings as *ideas*, and to express those ideas as words, introduces an entirely new dimension into the universe.

First, language – and it is a most extraordinary thing – allows us to 'think', by assigning words to objects and ideas. Then it becomes possible to express a *logical* idea by applying grammatical rules to the arrangement of those words, and linking them together sequentially in a sentence. And more, the faculty of language allows us to take those thoughts 'brought into existence' by language and insert them with complete precision into the minds of others for them to share, or to disagree with. Language makes the world intelligible, by allowing man to transmit his thoughts and experiences in the form of accumulated knowledge from generation to generation – leading, perhaps inevitably, to the moment at the close of the twentieth century when he would 'hold in his mind's eye' the history of the universe he inhabits. Language makes it possible to distinguish truth, the faithful reflection of reality, from falsehood, and this, as the philosopher Richard Swinburne points out, is the foundation of *reason* (obviously), but also of *morality*, for 'it gives man the capacity to contrast the worth of one action to that of another, to choose what he believes worthwhile . . . and that gives us a conception of the *goodness* of things'. Thus humans, like all living things, are biological beings constrained by nature's laws; nonetheless language liberates our mind from the confines of our material brain, allowing us to transcend time and space to explore the non-material world of thought, reason and emotion. So, 'All the major intellectual attributes of modern man are tied up in some way with language,' as Ian Tattersall argues. Where then did language come from?

The prevailing view, till recently, held that this remarkable faculty required no specific explanation, and could be readily accommodated within the standard evolutionary rubric of the transformation of the simple to the complex. Language is explained (or 'explained away') as an evolved form of communication, no different in principle from the grunts or calls of other species. 'I cannot doubt,' observed Darwin in *The Descent of Man*, 'that language owes its origin to the imitation and modification of various natural sounds, the voice of other animals and man's own instinctive cries…' So too contemporary evolutionary texts portray human language as an improved method of communication over that of our primate cousins, while emphasising the similarities in the larynx and vocal cords (which, however, are not so similar as they appear) as evidence for language's evolutionary origin. 'Language evolved to enable humans to exchange information,' observes Robin Dunbar of the University of Liverpool.

In the 1950s the famous linguist Noam Chomsky challenged this interpretation of language as a more sophisticated form of primate communication by drawing attention to the significance of the remarkable alacrity with which children learn to speak. Language flows so readily, a 'babbling' stream of feelings, thoughts and opinions filling every nook and cranny of our lives, it is easy to assume it must be simple, simple enough for children to 'pick up' as readily as they pick up measles. Prior to Chomsky, the standard view held that children learned to speak in the same way as blotting paper absorbs ink, by soaking up the words they heard and then reiterating them. Chomsky argued this could not be so, pointing out the skill with which very young children learn to speak lies far beyond the intellectual competency of their years, for while they must struggle to grasp the elementary principles of mathematics, they acquire language with astonishing ease. An infant starting from a situation not dissimilar to that of an adult in a room of foreigners all jabbering away incomprehensibly, nonetheless:

> 'Within a short span of time and with almost no direct instruction will have dissected the language into its minimal separable units of sound and meaning,' writes linguist Breyne Moskowitz. 'He will have discovered the rules of recombining sounds into words and recombining the words into meaningful sentences. Ten linguists working full-time for a decade analysing the structure of the English language could not programme a computer with a five-year-old's ability for language.'

The aptitude of the young mind in mastering the staggering complexity of language presupposed, Chomsky argued, that humans must possess some form of highly specific 'Language Acquisition Device' hardwired into their brains that somehow 'knows' the meaning of words and the grammatical forms necessary to make sense of them. How, otherwise, can an infant know when its mother says, 'Look at the cat!' that she is referring to the furry four-legged creature, and not to its whiskers, or the milk it is drinking. Further, the 'device' must not just know what is being referred to, but the grammatical rules that permit the same 'idea' expressed in different ways to have the same meaning ('John saw Mary' conveys the same message as 'Mary was seen by John'), but excluding meaningless variations. Further again, it transpires that children learn language in the same way, whether brought up in New Jersey or New Guinea, and acquire the same grammatical rules of 'present', 'past' and 'future' in the same sequence. This implies that the 'device' in turn must be sensitive to a Universal Grammar, common to all languages, which can pick up on the subtlest distinction of meaning.

Now, our primate cousins do not possess this 'device', which is why, clever as they are, they remain (in the words of the veteran chimpanzee-watcher Jane Goodall) 'trapped within themselves'. By contrast, every human society, no matter how 'primitive', has a language capable of 'expressing abstract concepts and complex trains of reasoning'. The million Stone Age inhabitants of the highlands of New Guinea, 'discovered' in 1930 after being cut off from the rest of the world for several thousands of years, spoke between them *eight hundred* different languages, each with its complex rules of syntax and grammar.

How then did the faculty of language come to colonise the human brain? 'There must have been a series of steps leading from no language at all to language as we now find it,' writes the linguist Steven Pinker, 'each step small enough to have been produced by random mutation [of genes] and with each intermediate grammar being useful to its possessor.' It is, of course, possible to imagine how language might have evolved in this way from a simpler form of communication or 'protolanguage', starting perhaps with gestures, moving on to simple words or phrases with a single meaning, with the rules for linking words into sentences coming later. Pinker's intended parallel between the means by which our species acquired language and the infant's rapid progress from *burbling* through *words* to *sentences* might seem plausible, in the way of all evolutionary explanations, and would indeed be reasonable if language simply 'facilitated the exchange of information'. But, as Chomsky pointed out so persuasively, language is also an autonomous, independent set of rules and meanings that impose order, make sense of the world 'out there'. Rules and meanings cannot evolve from the simple to the complex, they just 'are'. The structure of sentences is either meaning*ful* or meaning*less*. The naming of an object is either 'right' or 'wrong'. An elephant is an elephant, and not an anteater. Hence Chomsky insisted, against Pinker, that those seeking a scientific explanation for language could, if they so wished, describe it as having evolved 'so long as they realise that there is no substance for this assertion, that it amounts to nothing more than a *belief*'. This, of course, is no trivial controversy, for language is so intimately caught up in every aspect of 'being human' that to concede that it falls outside the conventional rubric of evolutionary explanation would be to concede that so does man.

The dispute over the evolutionary (or otherwise) origin of language remained irresoluble till the late 1980s, when the first PET scans revealed how the simplest of linguistic tasks involves vast tracts of the brain in a way that defies any simple scientific explanation. Here every mode of language, *thinking* about words, *reading* and *speaking* them, is represented in different parts. The prosaic task of associating the word 'chair' with 'sit' generates a blizzard of electrical activity across the whole surface of the brain. Further, those scanning investigations revealed how, in the twinkling of a second that it takes to speak or hear a word, the brain fragments it into its constituent parts through four distinct modules concerned with spelling, sound (phonology), meaning (semantics) and articulation. These 'modules' are in turn then further subdivided *ad* (virtually) *infinitum*. The perception of sound, for example discriminating between the consonants 'P' and 'B', is represented in twenty-two sites scattered throughout the brain. There is something absolutely awe-inspiring in supposing we understand a word like 'elephant' only after it has been parsed in this way. And then, to compound it all, the brain must simultaneously while 'parsing' elephant also comprehend its meaning in its entirety, for the constituent symbols can really only be understood within the context of the whole word.

Naming Letters

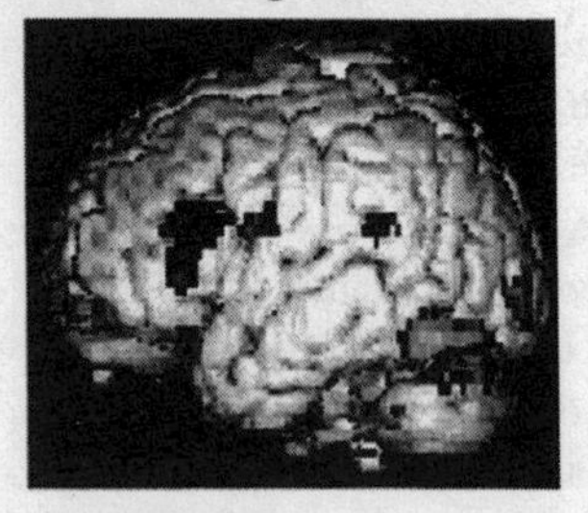

Naming Colours

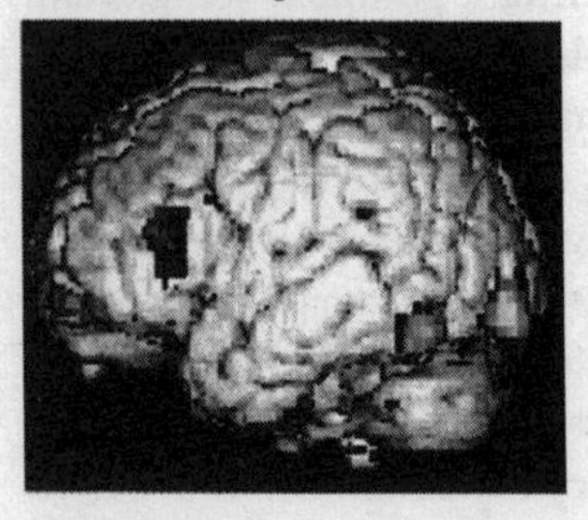

Repeating Heard Words

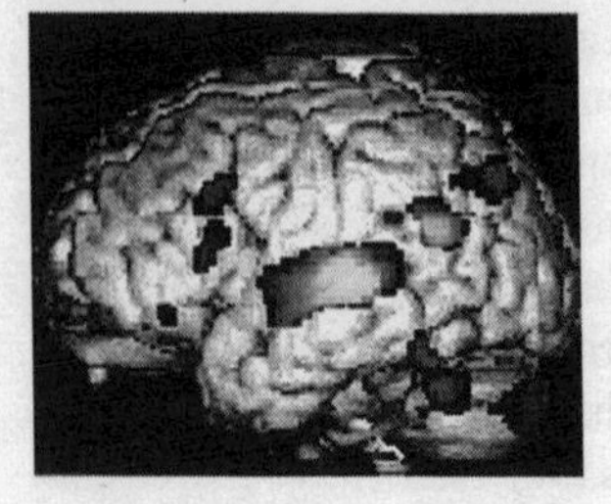

Self-Generated Words

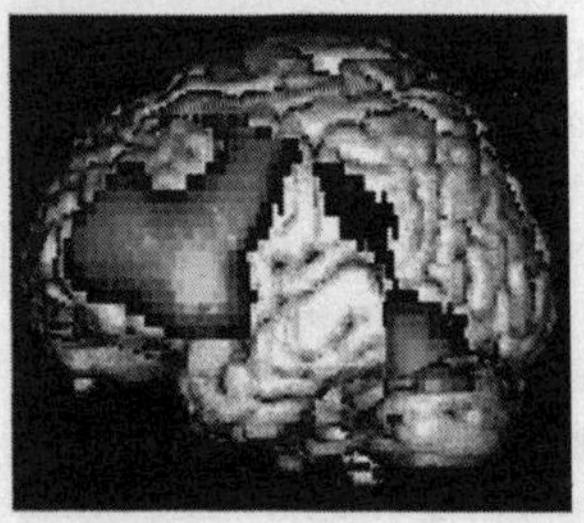

Fig 2-18: *The faculty of language, 'a complex system of knowledge about words', is comprised of half a dozen 'subsystems' processed in different parts of the brain – as shown by these brain-scanning studies of subjects naming letters and colours and repeating words, while spontaneous speech ('self-generated words') activates vast tracts of the frontal and parietal lobes.*

It is one thing to try to work out how the brain processes a single word (and that is baffling enough), quite another to extrapolate from such findings to try to imagine the same processes as they apply to a sentence, with its structure of 'subject-verb-object' and numerous subsidiary clauses. Move into the real world, with its ceaseless conversation, and the problem becomes insuperable. What sort of brain processes, one might ask, must be involved when a group of football fans convening in the pub before a match discuss their team's prospects for the coming season – drawing on a vast storehouse of knowledge and judgement of the form of previous seasons, the strengths and weaknesses of their players, and assessments of the performance of their rivals? How do they pluck from the storehouse of their memories the right words, or conjure from the rules of syntax and grammar the correct sequence with which emphatically to argue their opinion? How does the electrical firing of the neurons in their brains represent words and capture the nuance of their meanings?

And so? 'Language flows so readily, that it is easy to assume it must be simple.' But language only appears simple because it *has* to be so. There would, after all, be little point in humans acquiring this novel and powerful mode of inserting their thoughts directly into the minds of others if it took many years to get the hang of, and was difficult to use. But that apparent simplicity is, as already noted, a mark of language's profundity, concealing the inscrutable complexities of brain function that make it appear to be so.

The major legacy of linguistics and neuroscience in the past few decades has been to reveal the complexities concealed behind that apparent simplicity while drawing attention at the same time to how the faculty of language requires major changes in every aspect of the functioning of the brain: a massive increase in its memory capacity so as to be able to store that vocabulary of forty thousand words, together with the provision for their near-instant recall; a profound deepening of the mind's emotional repertoire with its feelings of sympathy and affection; the powers of reason; the moral distinction between right and wrong; and the imaginative intelligence with which poets and writers express themselves in unique ways.

The opportunity to reflect further on such matters will come later, but for the moment we must briefly return to contrast the conventional evolutionary portrayal of the origins of that 'totally unprecedented entity' Cromagnon man with how, in the light of the above, they now appear to be. To be sure, that steadily expanding brain over the preceding several million years, with its much enhanced neuronal firepower predisposed to those higher intellectual attributes, particularly language, and thus that cultural explosion of technical innovations and artistic expression. But that much-expanded brain by itself does not explain the phenomenon of language, nor why the evidence for its undoubted 'benefit' of being able to think, act and make sense of the world should have emerged so late and so suddenly. Why did the brain continue to expand in size for those millions of years when the 'pay-off' was so slight, and the attendant hazards of obstructed labour and dependent offspring so large? And this conundrum becomes yet more puzzling now we know that language is not just some bolt-on addition to the primate brain, but occupies large areas of it, and required the massive extension of those other attributes of mind, such as memory and intelligence, on which it depends.

Here neither of the two proposed evolutionary scenarios – that language evolved 'early' or 'late' – is convincing. The proponents of the 'early' scenario infer (quite rightly) that it must have taken millions of years for so complex a system to have evolved – all the way back to Turkana Boy's people, *Homo erectus*, and beyond. Why then, one might ask, did he exhibit so little evidence of the 'culture' that language makes possible? The 'late' theorists claim language to be unique to *Homo sapiens*, the spark that lit the cultural explosion that separates him from his nearest relatives – but that would presuppose that it evolved over the mere 100,000 years since his emergence from Africa. This dispute cannot be resolved, but it serves the useful purpose of drawing attention to our profound ignorance: we no longer have the vaguest inkling of what caused the 'switch to be thrown' to inaugurate that first and most astonishing of all civilisations. Thirty-five thousand years on, we humans can draw on a vast treasure house of the cumulative knowledge and technology of the many civilisations that have had their moment in the sun, the Egyptians, Grecks, Romans, Arabs and so on. The genius of the Cromagnons, with their passion for art and wittily decorated spear-throwers, is that they had to work it all out for themselves.

This then is the riddle of the Ascent of Man: how and why twenty or more distinct species of hominid should, over a period of several million years, have undergone that wholescale anatomical transformation required for standing upright, and then followed it up with acquiring that prodigiously sized brain whose potential to comprehend the workings of the universe appears so disproportionate to the needs of the life of a hunter gatherer. It seems obvious that man's sophistication and intelligence would have conferred some biological advantage, but all living things – birds, bats, dolphins and so on – have their own highly specialised sort of intelligence, different from our own, but which nonetheless maximises their chances of survival. The question, rather, as the biologist Robert Wesson puts it, is why the human brain should come with those striking mental powers, such as the capacity to compose symphonies or to solve abstruse mathematical theorems, that 'are not of the kind likely to be rewarded by numbers of descendants'.

The further subsidiary and related riddle is why, for the best part of 150 years, the scientific orthodoxy has prevailed that we know the answer, at least in principle, to that riddle of the Ascent, when, as the palaeontologist Ian Tattersall acknowledges, 'we have only the dimmest perception of how that dramatic history unfolded'. It has taken just a few pages to draw out the contradictions, at every turn, in the prevailing scientific certainty of 'natural selection' as the driving force of the Ascent of Man. There is, of course, no more self-evident truism than that nature 'selects' the strong and the fit at the expense of the weak and less than perfect. But that mechanism, by the same logic, can scarcely be invoked to account for standing upright and that massively enlarged brain which, by rights, should have so gravely compromised the survival prospects of those distant ancestors. There is nothing obscure in the observations outlined above: the anatomical implications of the upright stance and the obstetric hazards of that enlarging brain are well documented. Yet there is not the slightest hint in standard evolutionary texts or in the graphic museum displays of the Ascent that they might be problematic – while those who might think so are derisively dismissed, as we have seen, as 'stupid, ignorant or insane'.

Most people get by well enough without the slightest inclination to speculate about their origins – and if they do, there is much consolation in that reassuring image first captured by Thomas Huxley of our onward and upward ascent. Still, it is surprising how that history of our origins becomes instantly so much more fascinating and intriguing the moment one reflects, for example, on the marvels of the composite integrity of the human skeleton, or the hidden complexities of grammar that can nonetheless be grasped by a two-year-old child. This discrepancy between the beguiling simplicities of the evolutionary theory and the profundity of the biological phenomena it seeks to explain is very striking. Its claims can never be 'put to the test' of experimental verification, as there is no way of telling one way or the other whether the process of natural selection really does account for those extraordinary biological events millions of years ago. The standard evolutionary explanation is, in short, irrefutable – or *was* irrefutable, until the uncompromising verdict of the genome projects, where the random genetic mutations that might set us apart from our primate cousins, mice, flies or worms are nowhere to be found.

It can, admittedly, be very difficult to see what all this might add up to, but clearly the ramifications of those seemingly 'disappointing' outcomes of the New Genetics and the Decade of the Brain run very deep indeed. We need to know why we have been seduced into supposing that science knows so much more than is clearly the case – and that means exploring further that seemingly unbridgeable gap between those two 'Orders of Reality' to seek out the forces that might conjure the beauty and complexity of the natural world from the monotony of those chemical genes, and the richness of human thought and imagination from the electrical activity of the brain.

But for that we need a yet broader, more Olympian perspective still, to take the full measure of the scope (and limits) of scientific knowledge as so recently revealed, and of how, paradoxically, those 'disappointing' outcomes turn out to reveal profound truths about the nature of genetic inheritance and the human mind, so long concealed from view.

There is no better way to start than through that most fruitful insight into the nature of things that comes with the experience of 'wonder', whose dual meaning those Cromagnons would instinctively have appreciated. They would have 'wondered *at*' the pervasive beauty and integrity of the natural world, inferring there was a greater significance to their existence than they could know. They would have responded, too, to the human imperative to 'wonder *why*', seeking out in the regularity of the movement of the stars and the diversity of form of living things those causes, patterns and explanations of the natural world that are 'the beginning of all knowledge'.

3

The Limits of Science 1: The Quixotic Universe

'The world will never starve for want of wonders, but only for want of wonder.'

G.K. Chesterton

The world is so full of wonder, it is a wonder we do not see it to be more so. Every dawn the 'undeviating and punctual' sun rises on the horizon to flood our lives with the light and warmth that drive the great cycle of organic life – thirty million times more natural energy in a single second than that generated by manmade power stations in a whole year. And punctually at dusk, its setting brings the day to a close with a triumphant explosion of purple, red and orange streaked across the sky. 'Of all the gifts bestowed upon us,' wrote the Victorian art critic John Ruskin, 'colour is the holiest, the most divine, the most solemn.' Those limitless nuances of colour and light that suffuse our daily lives mark too the procession of the seasons, a constant reminder of the profound mystery of self-renewing life.

And there is nothing so full of wonder as life itself, the more so now we know that the vital actions of even the humblest bacterium, smaller by far than the full stop at the end of this sentence, involve the concerted action of thousands of separate chemical reactions, by which it transforms the nutrients absorbed from soil and water into the energy and raw materials with which it grows and reproduces itself. But life there is, and marching down through the ages in such an abundance of diversity of shape, form, attributes and propensities as to encompass the full range and more of what might be possible. And what variety!

'No one can say just how many species there are in these greenhouse-humid jungles,' writes naturalist and broadcaster David Attenborough of the forests of South America.

> There are over forty different species of parrot, over seventy different [species of] monkeys, three hundred [species of] humming birds and tens of thousands of [species of] butterflies. If you are not careful, you can even be bitten by a hundred different kinds of mosquito . . . Spend a day in the forest, turning over logs, looking beneath bark, sifting through the moist litter of leaves and you will collect hundreds of different kinds of small creatures: moths, caterpillars, spiders, long-nosed bugs, luminous beetles, harmless butterflies disguised as wasps, wasps shaped like ants, sticks that walk, leaves that open wings and fly . . . One of these creatures at least will almost certainly be undescribed by science.

And the millions of species with which we share the planet themselves represent a mere 1 per cent of those that have ever been, each form of life the opportunity for a further myriad of subtly different variations on a theme. Why should the extraordinary faces of the bat family, whose near-blindness should make them indifferent to physical appearances, nonetheless exhaust the possibilities of the design in the detailed geometry of their faces? Why should the many thousands of species of birds yet be so readily distinguishable one from the other by their pattern of flight or the shape of their wing, the colour of their plumage or the notes of their song?

But birds, as the American naturalist Frank Chapman once observed, are 'nature's highest expression of beauty, joy and truth', whose annual migration exemplifies that further recurring mystery of the biological world, those idiosyncrasies of habits and behaviour that defy all reason – like the Arctic tern, that every year traverses the globe, setting out from its nesting grounds in northern Canada and Siberia, winging its way down the coasts of Europe and Africa to the shores of the Antarctic, only to turn round and return northwards again: a round journey of twenty-five thousand miles that takes them eight months, flying twenty-four hours a day. How swiftly they fly, how confidently across the pathless sea at night!

And while we might rightly wonder how the Arctic tern knows how to navigate by the stars, it seems almost more wonderful still that the salmon should find its way from the depths of the ocean back to the

Fig 3-1: *Nature's fantasies. The bizarrely-shaped faces of the more than a thousand species of bat 'exhaust the possibilities of design', their prominent ears compensating for their poor vision by recapturing the sound waves they emit from their vocal cords as they rebound from their prey.*

same small stream from whence it set out, detecting through its highly developed sense of smell the waters of its spawning ground; or that the common European eel should cross the Atlantic twice, first from its breeding grounds off the North American coast to the rivers of Europe – and then back again. 'The number of [such] admirable, more or less inexplicable traits that one might cite is limited not by the inventiveness of nature,' writes biologist Robert Wesson, 'but rather by the ability of scientists to describe them.' There are, he points out, an estimated twenty thousand species of ant, of which only eight thousand have been described. So far biologists have got round to studying just one hundred of them in depth, each of which has its own unique, bizarre pattern of behaviour – such as 'the female of a parasitic ant which on finding a colony of its host, seizes a worker, rubs it with brushes on her legs to transfer its scent making her acceptable to enter the host colony'. How did there come to be such sophisticated and purposive patterns of behaviour in such minute creatures?

And yet that near-infinite diversity of life is permeated by an underlying unity, where everything connects in the same web of self-renewing life. The rain falling on the mountains feeds the springs that fill the streams. Those streams become rivers and flow to the sea, the mists rise from the deep and clouds are formed, which break again as rain on the mountainside. The plants on that mountainside capture the rainwater and, warmed by the energy of the sun, transform the nutrients of the soil, by some extraordinary alchemy, into themselves. A grazing animal eats that same plant to set up another complex web of connections, for it in turn is eaten by another, and its remains will return to the earth, where the microbes in the soil cannibalise its bones, turning them back into their constituent chemicals. And so the process of reincarnation continues. Nothing is lost, but nothing stays the same.

Wheels within wheels; and across that vast landscape of living things, from the highest to the lowest, the survival and prosperity of man is yet, as J. Arthur Thomson, Professor of Natural History at Aberdeen University reminds us, completely dependent on the labours of the humble earthworm, without whose exertions in aerating the dense, inhospitable soil there could never have been a single field of corn.

> When we pause to think of the part earthworms have played in the history of the earth, they are clearly *the* most useful of animals. By their burrowing, they loosen the earth, making way for the plant rootlets and the raindrops; by bruising the soil in their gizzards they reduce the mineral particles to more useful forms; they were ploughers before the plough. Five hundred thousand to an acre passing ten tons of soil every year through their bodies.

So, the world 'will never starve for want of wonders', the more so for knowing and wondering how the sky above and the earth below and 'all that dwell therein' – including the human mind, with its powers of reason and imagination – originated as a mass of formless atoms in that 'moment of singularity' of the Big Bang fifteen billion years ago.

The poet William Wordsworth, seeking to catch the enfolding delight of that sky above and earth below, called it 'the sublime',

> Whose dwelling is the light of setting suns,
> And the round ocean and the living air,
> And the blue sky,
> A spirit that impels and rolls through all things.

The feelings evoked by nature and 'the sublime' were, for the American poet Walt Whitman as for so many poets and writers, the most powerful evidence for a hidden, mystical core to everyday reality.

> 'There is, apart from mere intellect,' he wrote, 'a wondrous *something* that realises without argument an intuition of the absolute balance, in time and space of the whole of this multifariousness we call *the world*; a sight of that unseen thread which holds all history and time, and all events like a leashed dog in the hand of the hunter.'

That sublime nature has always provided the most powerful impetus to the religious view, its celebration a central feature of all the great religions. For the German theologian Rudolph Otto (1869–1937), the 'sublime' was a '*mysterium tremendum et fascinans*': both *awesome*, in whose presence we feel something much greater than our insignificant selves, and also *fascinating*, compelling the human mind to investigate its fundamental laws.

This brings us to the second of the dual meanings of 'wonder' suggested at the close of the preceding chapter, to 'wonder *why*', which, as the Greek philosopher Plato observed, 'is the beginning of all knowledge'.

> 'The scientist does not study nature because it is useful to do so,' wrote the nineteenth-century French mathematician Henri Poincaré. 'He studies it because he takes pleasure in it; and he takes pleasure in it because it is beautiful. If nature were not beautiful, it would not be worth knowing, and life would not be worth living . . . I mean the intimate beauty which comes from the harmonious order of its parts and which a pure intelligence can grasp.'

The greatest (probably) of all scientists, Isaac Newton, seeking to comprehend that 'harmonious order of parts', would discover the fundamental laws of gravity and motion, that, being Universal (they hold throughout the universe), Absolute (unchallengeable), Eternal (holding for all time) and Omnipotent (all-powerful), he inferred, offered a glimpse into the mind of the Creator. Newton captured this dual meaning of wonder, to 'wonder at' and to 'wonder why', in his famous confession that the most he could hope to achieve was to illuminate the workings of some small part of that sublime world: 'I do not know what I may appear to the world,' he wrote, 'but to myself I seem to have been only like a boy playing on the sea shore, diverting myself now and then, finding a smoother pebble than ordinary, whilst the great ocean of truth lay all undiscovered before me.'

The wonders of the world are so pervasive that to the seemingly less sophisticated minds of earlier ages (such as Newton's) they were best understood as 'natural miracles'. To be sure, the undeviating and punctual sun, the cycle of life, the infinite variety of living things, their interconnectedness to each other, these are all part of nature, and are faithful to its laws. They are 'natural'. But the totality of it all, its beauty and integrity and completeness, that 'great undiscovered ocean of truth', lie so far beyond the power of the human mind to properly comprehend, they might as well be 'a miracle'. Thus science and religion were cheerfully reconciled, the scientist seeing his task as a holy calling, where Robert Boyle, the founder of modern physics, would perceive his role as 'a priest in the temple of nature'.

This is scarcely the modern view. Most people, of course, acknowledge the beauty and complexity of the world and find it admirable, even uplifting – but you could search in vain for a textbook of biology or zoology, astronomy or botany, or indeed of any scientific discipline, which even hints that there is something astonishing, extraordinary, let alone 'miraculous', about its subject. Science no longer 'does' wonder, which is more readily associated nowadays with the incurious mysticism and incense of the New Age. Science prefers to cultivate an aura of intellectual neutrality, the better to convey its disinterested objectivity, its commitment to the 'truth'. Hence the highly technical, and to the outsider often impenetrable, prose of its texts and learned journals, from which any sense of wonder is rigorously excluded.

There are, as will be seen, several important reasons for this modern-day lack of astonishment, but the most important is undoubtedly the general perception that science, since Newton's time, has revealed those 'natural miracles' to have a distinctly non-miraculous, materialist explanation – culminating in that firestorm of scientific discovery of the past fifty years, which has integrated into one coherent narrative the entire history of the universe from its origins to the present day. To be sure, science may not capture the beauty and connectedness of it all, the 'sublime spirit that rolls through all things', but this is more than compensated for by the sheer drama and excitement of the events it has so convincingly described.

The scale of that intellectual achievement is so great that there might seem little room any more for the 'natural miracles' of an earlier age, or to 'wonder' whether there might after all be more than we can know. It would certainly require a truly Olympian perspective, capable of surveying the vast landscape of science, to recognise where and what the limits to its knowledge might be – and that would seem an impossibility. Yet it is not quite so, for while that landscape is indeed vast, and far beyond the comprehending of any individual, it is nonetheless sustained by three great unifying phenomena that impose *order* on the world – which on examination can tell us something very profound about science and the limits of its materialist explanations.

It is fruitless – always has been, always will be – to pose that most elementary of all questions: 'Why is there something rather than nothing?' The same however does not apply to the second and supplementary question: 'Why, *given there is something*, are both the physical universe (and all that it contains) and all life (in its infinite diversity) so *ordered*?' They should not be, for anything left to itself will tend towards chaos and disorder, as fires burn out and clocks run down – unless countered by a compensating force imposing order, restituting lost energy.

There are (to put it simply) three 'forces for order': first the force of *gravity*, as discovered by Sir Isaac Newton, the glue that binds the universe together; next the all-powerful genes strung out along the *Double Helix*, imposing the order of *form*, the shape, characteristics and attibutes unique to each of the millions of species of living things; and thirdly the *human mind*, that imposes the order of *understanding* on the natural world and our place within it. These three forces control or sustain all (or virtually all) phenomena in the universe, and stand proxy for the 'vast landscape' of science. Thus, if they are knowable scientifically as belonging to that materialist, second-order reality of the physics and chemistry of matter (where water is a combination of two atoms of hydrogen and one of oxygen), then by definition there is nothing in theory that science cannot know. But if they are not so knowable, one can only infer that they exert their effects through some other force that lies beyond the range of science and its methods to detect. We start with Sir Isaac Newton's theory of gravity.

Fig 3-2: *Sir Isaac Newton in his* Philosophiae Naturalis Principia Mathematica, *'the greatest single work in the history of science', described the three laws of motion and universal gravitation. Here, his profile in the centre of a starry firmament, with allegorical figures and scientific instruments, reflects his near-mythical status.*

Isaac Newton, born in 1642 into a semi-literate sheep-farming family in rural Lincolnshire, was one of the tiny handful of supreme geniuses who have shaped the categories of human knowledge. From the time of Aristotle onwards, and for the best part of two thousand years, the regularity and order of the physical world was as it was because it was divinely ordained to be so: the punctual and undeviating sun, the movement of the planets across the heavens, the passage of the seasons and apples falling from trees. Newton's genius was to realise that these and numerous other aspects of the physical world were all linked together by the hidden force of gravity.

Soon after graduating at the age of twenty-three from Cambridge University, Newton was compelled by an epidemic of bubonic plague to return to his home in Lincolnshire. There, over a period of just two years, he made a series of scientific discoveries that would not be equalled till Einstein, almost 250 years later. These included the nature of light and the mathematical method of differential calculus, with which it is possible to calculate the movement of the planets in their orbit. Newton's most famous insight came when, sitting in his garden, he saw an apple fall from a tree. He 'wondered' whether the force of the earth's gravity pulling the apple to the ground might reach still further, and hold the moon in its orbit around the earth.

Newton's friend Dr William Stukeley would later record his reminiscences of that great moment.

> After dinner, the weather being warm, we went into the garden and drank tea, under the shade of some apple trees, only he and myself. Amidst other discourse, he told me he was just in the same situation as when, formerly, the notion of gravitation came into his mind. It was occasioned by the fall of an apple, as he sat in a contemplative mood. Why should that apple always descend perpendicularly to the ground, thought he to himself? Why should it not go sideways or upwards, but constantly to the earth's centre? Assuredly, the reason is, that the earth draws it. There must be a drawing power in matter ... and if matter thus draws matter, it ... must extend itself through the universe.

Newton's 'notion of gravitation', of 'matter drawing on matter', would resolve the greatest conundrum of the movement of those heavenly bodies, why they remained in their stately orbits (the moon around the earth, the earth around the sun) rather than, as they should by rights, being impelled by their centrifugal force into the far depths of outer space. Newton, being a mathematical genius, calculated the strength of that countervailing force of gravity, showing it to be determined by the masses of the moon and earth, earth and sun respectively, multiplied together and divided by the square of the the distance between them, and so too throughout the entire universe. By the time Newton published his epic three-volume *Principia Mathematica* in 1697, describing the theory of gravity and the three laws of motion, he had transformed the divinely ordained physical world into which he was born into one governed by absolute and unchallengeable universal laws known to man, where everything was linked to everything else in a never-ending series of causes – all the way into the past and indefinitely into the future.

From the beginning, the force of gravity at the moment of the Big Bang imposed the necessary order on those billions of elementary particles, concentrating them into massive, heat-generating stars. Several thousand millions of years later, the same force of gravity would impose order on our solar system, concentrating 99 per cent of its matter within the sun to generate the prodigious amounts of energy, heat and light that would allow the emergence of life on earth. And anticipating the future? Newton's friend, the Astronomer Royal Edmond Halley, used Newton's laws to work out the elliptical orbit of the comet that bears his name and so predict its seventy-six-year cycle of return. Three hundred years later, NASA scientists would use those same laws to plot the trajectory of the first manned space flight to the moon. And it is even possible to predict when it will all end – in five thousand million years' time (or thereabouts), when the prodigious energy generated by our sun will be exhausted, and our earth will perish.

As time has passed, so the explanatory power of Newton's laws of gravity has grown ever wider, to touch virtually every aspect of human experience: the movement of the sun and stars (obviously), the waxing and waning of the moon, the ebb and flow of the tides, the contrasting climates of the Arctic Circle and the sand-swept desert, the cycle of the seasons, rain falling on the ground, the shape of mountains sculpted by the movement of glaciers, the flow of rivers towards the sea, the size of living things from whale to flea and indeed ourselves – for we could not be any bigger than we are without encountering the hazard, posed by gravity, of falling over.

Newton's laws epitomise, to the highest degree, the explanatory power of science, through which for the first time we humans could comprehend the workings of that vast universe to which we belong. But, and it is a most extraordinary thing, three hundred years on, the means by which the powerful, invisible glue of gravity imposes order on the universe remains quite unknown. Consider, by analogy, a child whirling a ball attached to a string around its head, just as gravity holds the earth in its perpetual orbit around the sun. Here, the string (like gravity) counteracts the centrifugal force that would hurtle the ball (the earth) into a distant tree. *But there is no string*. Newton himself was only too well aware that there had to be some physical means by which gravity must exert its influence over hundreds, thousands, of millions of miles of empty space. It was, he wrote, 'an absurdity that no thinking man can ever fall into' to suppose that gravity 'could act at a distance through a vacuum without the mediation of anything else, by which that action and force may be conveyed'.

Perhaps, he speculated, space was suffused by an invisible 'ether' composed of very small particles that repelled one another and by which the sun could hold the earth in its orbit – though this would mean that over a very long period the movement of the planets would gradually slow down through the effects of friction. But in 1887 the American physicists, Albert Michelson and Edward Morley, discovered that there was no 'ether'. Space is well named – it is empty. Put another way, Newton's theory encompasses the profound contradiction of gravity being both an immensely powerful force imposing order on the matter of the universe, linking its history all the way back to the beginning and anticipating its end, yet itself being *non*-material. This extraordinary property of gravitation requires some sort of context, by contrasting it with, for example, that equally potent invisible 'force' electricity, which at the touch of a switch floods the room with light. But whereas electricity is a 'material' force – the vibration of electrons passing along a copper wire – gravity exerts its effects across billions of miles of empty space, through a vacuum of nothingness.

Newton's theory stands (for all time), but has been modified in two directions. First, in 1915, Albert Einstein in his General Theory of Relativity reformulated the concept of gravity to allow for space to be 'elastic', so that a star like our sun could curve and stretch the space around it – and the bigger the star, the greater the effect. Matter, Einstein showed, warps space. This takes care of the more bizarre phenomena in the universe, such as 'black holes', that capture even the weightless particles of light – but for all that the profound Newtonian mystery of how gravity exerts its force through the vacuum of space remains unresolved.

Next, it has emerged that Newton's gravitational force is not alone, being just one of four (similarly non-material) forces, including those that bind together the atomic particles of protons and neutrons – whose disruption generates the prodigious energy of an atomic explosion. In the twentieth century, the conundrum of the non-materiality of those gravitational forces was compounded when it emerged that their strength is precisely tuned to permit the consequent emergence of life and ourselves. If the force they exert were, for example, ever so slightly *stronger*, then stars (like our sun) would attract more matter from interstellar space, and being so much bigger would burn much more rapidly and intensely – just as a large bonfire outburns a smaller one. They would then exhaust themselves in as little as ten million years, instead of the several billion necessary for life to 'get going'. If, contrariwise, the force of gravity were ever so slightly *weaker*, the reverse would apply, and the sun and stars would not be big enough to generate those prodigious amounts of heat and energy. The sky would be empty at night, and once again we humans would never have been around to appreciate it. It is, of course, very difficult to convey just how precise those forces necessary for the creation of the universe (and the subsequent emergence of life on planet earth) had to be, but physicist John Polkinghorne estimates their fine tuning had to be accurate to within one part in a trillion trillion (and several trillion more), a figure greater by far than all the particles in the universe – a degree of accuracy, it is estimated, equivalent to hitting a target an inch wide on the other side of the universe.

Isaac Newton's theory of gravity is the most elegant idea in the history of science. Nothing touches its combination of pure simplicity, readily understandable by a class of ten-year-olds, and all-encompassing explanatory power. His contemporaries were dazzled that so elementary a mathematical formula could account for so much – prompting the poet Alexander Pope to propose as his epitaph in Westminster Abbey:

> Nature and nature's laws lay hidden in night
> God said *Let Newton be*! And all was light.

Still, Newton's gravitational force, imposing 'order' on the physical universe, clearly fails the test of scientific 'knowability', for while we can fully comprehend all its consequences we are 'left with that absurdity that no thinking man can ever fall into' of having to suppose that a *non*-material force can 'act at a distance' across millions of miles of empty space without the mediation of anything by which that action and force may be conveyed. Thus, ironically, this most scientific of theories, grounded in the observation of the movements of the planets expressed in mathematical form, subverts the scientific or materialist view which holds that everything must ultimately be explicable in terms of its material properties alone.

We turn now to the living world of plants, insects, fishes, birds and ourselves, which is billions upon billions upon billions of times more complex than Newton's non-living, physical universe. Hence, the two forces that impose order on that world, the Double Helix imposing the order of *form* on living things, and the human brain and its mind imposing the order of *understanding*, will be profounder than the glue of gravity by similar orders of magnitude. We might anticipate that these two further forces of order will, like Newton's theory of gravity, similarly prove to be non-material, and therefore fail the test of scientific knowability. But to 'get there' we must first come to grips with how we have come to suppose otherwise, and specifically how in the mid-nineteenth century Darwin's grand evolutionary theory, as set out in the twin texts of *On the Origin of Species* and *The Descent of Man*, offered an apparently all-encompassing and exclusively materialist explanation for the phenomena of life.

4

The (Evolutionary) 'Reason for Everything': Certainty

> 'There is grandeur in this view of life, with its several powers, having been originally breathed into a few forms or into one . . . from so simple a beginning endless forms most beautiful and most wonderful have been evolved.'
>
> *Charles Darwin,* On the Origin of Species *(1859)*

Charles Darwin, while a theology student at Cambridge University, developed a passion for beetles. 'Nothing gave me so much pleasure,' he would write in his autobiography, recalling how, 'as proof of my zeal':

> One day, on tearing off some old bark I saw two rare beetles and seized one in each hand; then I saw a third and new kind, which I could not bear to lose, so I put the one which I held in my right hand into my mouth. But alas! It ejected some intensely acrid fluid that burnt my tongue so I was forced to spit it out and so it was lost.

Darwin's zeal for beetles was quite unexceptional, for he was born into the Golden Age of Natural History, when the wonders of nature as revealed by science gripped the public imagination with an extraordinary intensity, while being also the most tangible evidence of a divinely ordained world. 'The naturalist . . . sees the beautiful connection that subsists throughout the whole scheme of animated nature,' observed the editor of the *Zoological Journal* of London. 'He traces . . . the mutual depending that convinces him nothing is made in vain.'

There seemed no limit to the new forms of 'animated nature' just waiting to be discovered. In 1771 the famed maritime explorer James Cook had returned from his epic three-year circumnavigation of the world 'laden with the greatest treasure of natural history that ever was brought into any country at one time': no fewer than 1,400 new plant species, more than a thousand new species of animals, two hundred fish and assorted molluscs, insects and marine creatures. For his friend the anatomist John Hunter, waiting for him as his ship anchored off Deal harbour, Cook had several unusual specimens to add to his famous collection: a striped polecat from the Cape of Good Hope, part of a giant squid, and a peculiar animal 'as large as a greyhound, of mouse colour and very swift', known in the Aboriginal dialect as a '*kangooroo*'.

The discovery of this exhilarating diversity of life extended beyond the living to the long-since extinct. For this, too, was the great period of geological discovery of the antiquity of the earth, the strata of whose rocks revealed fossilised bones and teeth so much larger than any previously encountered as to suggest that vast, fantastical creatures had roamed the surface of the earth millions of years before man.

The immediate fascination of natural history lay in the accurate description of that teeming variety of life, but beyond that there was every reason to suppose that comparing the anatomical structure and the behaviour of living organisms such as the polecat, squid and kangaroo would reveal the long-suspected hidden laws that link all 'animated nature' together. The search for those laws stretches back into antiquity, seeking first to explain the 'vitality' of the living, the heat, energy and movement that so readily distinguish it from the non-living, and that depart so promptly at the moment of death. The subtler, yet related, question concerned the nature of 'form', those elusive qualities of pattern and order that so clearly distinguish polecat, squid and kangaroo from each other, and the tissues of which they are made – as readily as a grand palace is distinguished from a humble factory, and from the bricks and mortar of which they are constructed. But the elusive 'form' of polecat and squid, unlike that of the palace or the factory, has the further extraordinary property of remaining constant throughout their lives, even though the 'bricks and mortar' from which they are fashioned are being constantly replaced and renewed. From the first natural historian, Aristotle, onwards, it was presumed that some organising principle, some 'formative impulse', must both determine and ensure that constancy of form.

The presiding genius of natural history, Baron Georges Cuvier (1769–1832), director of the Musée d'Histoire Naturelle in Paris, proposed two laws of that 'formative impulse', the laws of *similarity* (homology) and *correlation*. First, homology. Cuvier inferred from a detailed study of the ten thousand specimens in his collection that the diverse forms of animals each concealed an underlying 'unity of type', all being variations on the same 'blueprint': the wings of bird and bat, the paddle of a porpoise, the horse's legs and the human forearm were all constructed from the same bones, adapted to their 'way of life' – whether flying or swimming, running or grasping.

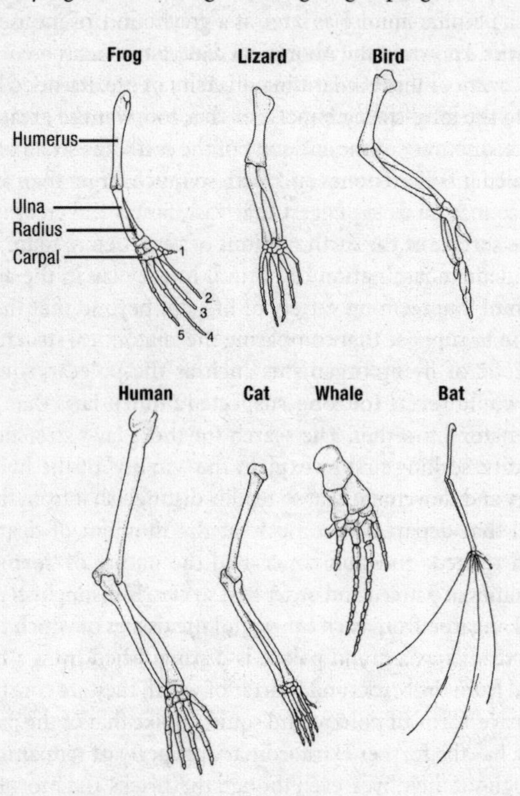

Fig 4-1: *Georges Cuvier's law of homology. The limbs of frog and lizard, bird and cat, whale, bat and man all show the same basic plan, being constructed from the same bones (humerus, radius, ulna, carpals, metacarpals, phalanges), modified for their 'way of life'.*

His second law, of 'correlation', asserted that the various parts of every animal, its skull, limbs, teeth, etc., were all 'of a piece', all correlated together, being so fashioned as to fulfil its way of life. Thus a carnivore, such as a lion or hyena, would have limbs strong enough to grasp its victim and muscular enough for hunting, jaws sufficiently powerful and teeth sharp enough to rip its flesh, and so on. 'Every organised being forms a whole, a unique and perfect system, the parts of which mutually correspond and concur in the same definitive action,' he wrote.

Cuvier maintained that these laws dictating the harmony of the parts of the 'unique and perfect system' were as precise as those of mathematics. He could not specify the biological forces behind them, but they were not merely some theoretical inference. Rather, they could be 'put to the test', allowing him, to the astonishment of all, to 'restore to life' those fantastical and long-extinct creatures from long ago, reconstructing from the assorted bones and teeth of their fossilised remains a 'megatherium', or 'huge beast', a creature resembling a giant sloth which would stand on two legs to graze on leaves. 'Is not Cuvier the great poet of our era?' enquired the novelist Honoré de Balzac. 'Our immortal naturalist has reconstructed past worlds from a few bleached bones . . . discovered a Giant population from the footprints of a mammoth.'

Fig 4-2: *The megatherium. In 1796 Cuvier reconstructed the skeleton of the megatherium from giant fossil remains found in Paraguay – one of the largest mammals ever to walk the earth, which standing on its hind legs was almost twice the height of an elephant.*

Cuvier's laws may have been firmly grounded in the scientific discipline of comparative anatomy, but they had obvious metaphysical implications. They implied, inescapably, some higher intelligence, responsible for drawing up the blueprint of the forelimbs and ensuring that the lion's jaws, teeth and muscles correlated so well together.

The theological implications of natural history were further elaborated, from a different perspective, by William Paley in his widely read *Natural Theology* (1802). Paley argued, and many were persuaded, that the beauty and perfection of living things offered the most persuasive evidence for there being a 'Grand Designer'. His central premise is summarised in the familiar opening paragraph, where he reflects on the significance of finding a watch on his path when 'crossing a heath':

> . . .suppose I pitch my foot against a *stone* and were asked how the stone came to be there; I might possibly answer that for anything I knew to the contrary it had lain there for ever . . . but suppose I had found a *watch* upon the ground . . . I should hardly think of the answer which I had given before . . . for this reason and for no other, viz, that when we come to inspect the watch we perceive (what we could not discover in the stone) that *its several parts are framed and put together for a purpose*, for instance they are so formed and adjusted to produce motion, and that motion so regulated as to point out the hour of the day.

He then proceeds to inspect the several parts of the watch, its elasticated spring, the series of wheels with their catching teeth, leading to the 'inevitable inference . . . that the watch must have had a maker'. And so too the natural world of plants, insects, birds, fish and mammals similarly shows 'every indication of contrivance, every manifestation of design', the only difference being that, when compared to a watch, the evidence of design 'is greater and more to a degree that exceeds all computation'.

Paley, in the first of many examples, compares the human eye with the telescope, in both of which the lens is designed for the same purpose – 'being adjusted to the laws by which rays of light are transmitted'. The further 'contrivances' of the eye necessary for vision, he points out, include the eyelids to protect and lubricate its cornea; the shutter-like iris, self-adjusting for intensity of light; the clear, jelly-like vitreous humour that sustains its shape; the ingenious muscles coordinating its

CHAPTER VI.

THE ARGUMENT CUMULATIVE.

Were there no example in the world of contrivance except that of the *eye*, it would be alone sufficient to support the conclusion which we draw from it, as to the necessity of an intelligent Creator. It could never be got rid of; because it could not be accounted for by any other supposition, which did not contradict all the principles we possess of knowledge; the principles according to which things do, as often as they can be brought to the test of experience, turn out to be true or false. Its coats and humours, con-

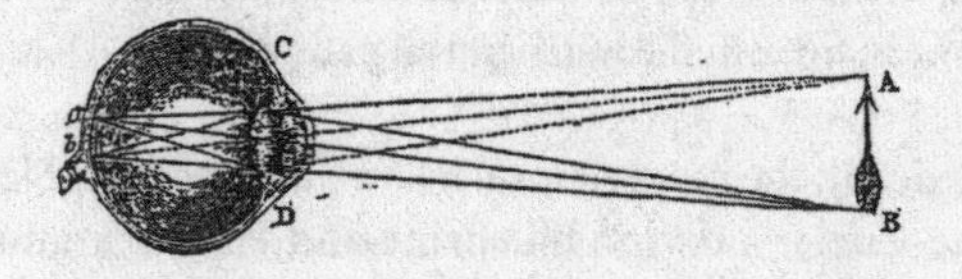

[The figure is introduced to remind the reader of the fine adjustment of the eye; a subject explained in the Appendix:—A, B, is the object, and the lines represent the light reflected from it into the eye. On the surface of the cornea, which is the transparent part of the eye,

Fig 4-3: *William Paley's 'Argument Cumulative'. The lens of the human eye, like that of a telescope, conforms to the optical laws of refraction by focusing the image on the back of the retina. For Paley, 'The eye alone would be sufficient to support the conclusion . . . as to the necessity of an intelligent creator.'*

movements; the optic nerve transmitting the electrical impulses to the brain. Each of these parts – lens, iris, retina, muscles and optic nerve – is 'perfectly designed' for its specific contribution to the overall purpose of seeing, but has no utility by itself, just as the springs of a watch are perfectly designed for their specific contribution to telling the time, but only within the context of the watch in its entirety.

Paley was a pioneer of the genre of 'popular science', bringing the most recent findings of comparative anatomy to a large audience, who would be deeply influenced by the seemingly irrefutable logic of his 'argument from design'. For thousands of years people had interpreted in a general way the wonders of the natural world as evidence for some higher intelligence. But Paley's demonstration of the perfection of design of the components of the eye, and their interdependence one with the other, presupposed some higher intelligence to have conceived of the eye in its totality, and to have ensured that its constituent parts were so designed as to make possible its purpose of seeing.

It is difficult to imagine nowadays what it must have meant to live in this divinely created world, where every aspect of nature spoke of God's 'wisdom and providence'.

> 'There came into my mind so sweet a sense of the glorious majesty and grace of God,' wrote the American theologian Jonathan Edwards, describing a walk in the New Jersey countryside, '*a divine glory in everything* – in the sun, moon and stars; in the clouds and blue sky; in the grass, flowers, trees; in the water and all nature.'

But paradoxically, the more natural history advanced, the less convincing became Cuvier's divine blueprint and Paley's argument for 'A Designer'. It seemed an absurdity, indeed demeaning, to suppose that God had nothing better to concern himself with than the minutiae of designing thousands of different species of insects – including, no doubt, those rare specimens of beetle that had attracted Darwin's attention. Natural history, in short, was in desperate need of a theory comparable to Newton's law of gravity, that would allow nature, governed by its own 'natural laws', to take care of itself. Perhaps, suggested the French naturalist Jean-Baptiste Lamarck, living things *designed themselves*, possessing some inherent facility to adapt to their particular environmental niche. So, the giraffe does not have a long neck because God had designed it to be so, but rather it might have extended its own neck, the better to graze off the leaves of the tops of trees. This 'acquired characteristic' would in turn be inherited by its progeny. So, 'Little by little,' Lamarck wrote, 'nature has succeeded in fashioning the animals such as we see before us.' His speculations were not well received, and, falling foul of the all-powerful Baron Cuvier, he died blind, penniless and unrecognised in 1829, to be buried in a pauper's grave. His notion of the possibility of the evolutionary transformation of species would, of course, live on.

The Origin of On the Origin

Meanwhile, Charles Darwin's enthusiasm for natural history had long since triumphed over his intended career in the Church, and in 1831, two years after Lamarck's death, he set sail from Plymouth on board HMS *Beagle* for its four-year circumnavigation of the world – with instructions to survey the shores of South America and 'some islands in the Pacific'. The winds of scientific optimism and theological doubt filled his sails. He was blessed too with the necessary virtues for a budding naturalist: a fascination with the living world, acute powers of observation, and a fine literary style that would ensure that those observations would be widely read. Three months into his journey he arrived off the coast of Brazil, and his delight in walking through the tropical forest for the first time bears comparison with the theologian Jonathan Edwards' 'divine glory in everything':

> The day has passed delightfully. Delight itself, however, is too weak a term to express the feelings of a naturalist who, for the first time, has wandered by himself in a Brazilian forest. The elegance of the grasses . . . the beauty of flowers, the glossy green of the foliage, but above all the general luxuriance of the vegetation filled me with admiration. To a person fond of natural history, such a day as this brings with it a deeper pleasure than he can ever hope to experience again.

Darwin's journal, subsequently published as *The Voyage of the Beagle*, captures the wonder of that world in the early nineteenth century, still for the most part unexplored, so full of the promise of new things to be discovered, so much vaster and more inaccessible than our own, its oceans wider, its mountains higher. It evokes too that romantic image of the traveller returning after a long absence and many adventures having 'made his fortune' – finding a goldmine in Ecuador or its equivalent. Darwin's bounty was of a different sort, a 'cataclysmic idea that would break up the crust of conventional opinion'.

The evolutionary theory for which he is famous emerged from the synthesis of two distinct patterns of nature: a geographical pattern in *space* – that is, the distribution of closely related but distinct species over large distances; and a historical pattern over *time*, linking the fossilised remains of extinct animals with those still living.

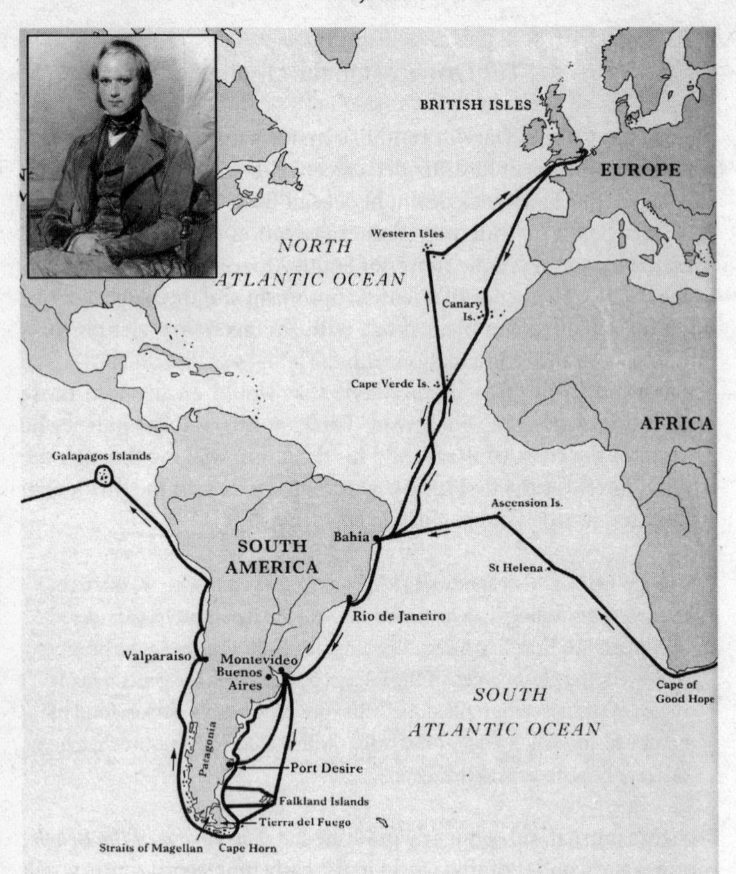

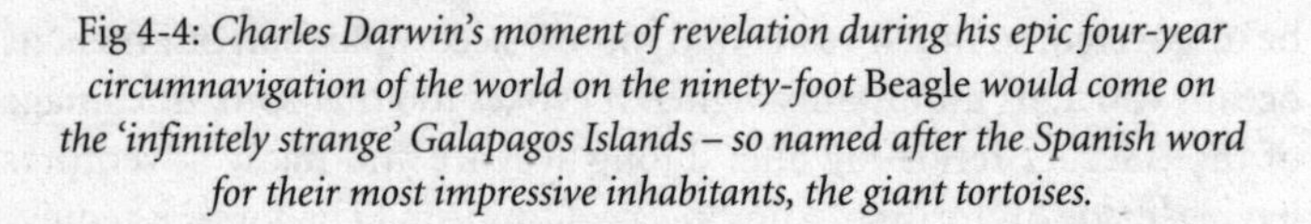

Fig 4-4: *Charles Darwin's moment of revelation during his epic four-year circumnavigation of the world on the ninety-foot* Beagle *would come on the 'infinitely strange' Galapagos Islands – so named after the Spanish word for their most impressive inhabitants, the giant tortoises.*

Sailing down the coast of South America in the *Beagle* and then back up the western seashore of Chile, Darwin noted a geographical pattern of similar, but distinct, species. He famously had his great moment of revelation on the 'infinitely strange' Galapagos Islands, six hundred miles off the coast of Ecuador, whose inhabitants included thousands of marine lizards, miniature dragons sunning themselves on the rocks, wingless cormorants and vast tortoises. Back on board, sorting out his specimens, he was struck by how the majority were unique to the Galapagos. Certainly they resembled those he had already observed on the mainland of South America, but at the same time they were quite different.

> 'It was most striking,' he would later write, 'to be surrounded by new birds, new reptiles, new insects, new plants, and yet by trifling details of structure and even by the tones of voice and plumage of the birds, to have the temperate plains of Patagonia, or the hot dry deserts of northern Chile, vividly brought before my eyes.'

He noted as the *Beagle* moved from island to island how each had its own species of lizard, tortoise and plants with the 'same general habits', but different in appearance one from the other. Most strikingly of all, he identified several different species of finch, each with its own particular-shaped beak, 'adapted to its particular method of finding food': one a powerful crushing nutcracker, another similar to a pair of probing tweezers (*see overleaf*). Clearly the Creator had been very busy indeed.

Paralleling that geographical variation in *space* were those in *time*. In Patagonia Darwin found a nearly entire skeleton of a bizarre horse-sized mammal with an enormous pelvis and a small, long face like an anteater. At another site he found the glyptodon, which bore a striking resemblance to the armadillo. 'Page after page of his notepad was filled up as he thought on the implications,' writes his biographer Adrian Desmond. 'What did the country look like in their day? And why had they all died out? He dreamed himself back to an archaic world where bull-sized sloths roamed unimaginable plains.'

Darwin returned from his epic voyage to find himself a minor celebrity, flattered by the decision of the exclusive Royal Society to elect him to their ranks. Yet his mind remained in limbo, seeking the thread that would link the observations of that epic journey together.

> 'The subject haunted me,' he would write in his autobiography. 'It was evident that such facts could only be explained on the supposition that species gradually became modified . . . But it was equally evident that neither the action of the surrounding conditions [i.e. the local environment], nor the will of organisms [a reference to Lamarck's theory

Fig 4-5: *Darwin's finches. The Galapagos finches differ so greatly in size, plumage, beak shape and behaviour as to be readily identified as distinct species. The variation in their beaks reflects the wide variety of their feeding habits. 'One might really fancy,' wrote Darwin, 'that one species had been taken and modified for different ends.'*

whereby the giraffe willed itself to have a longer neck] could account for the numerable cases in which organisms of every kind are beautifully adapted to their habits of life.'

Fifteen months after his return, Darwin had his own 'falling apple' experience when reading ('for amusement') the economist Thomas Malthus's *An Essay on the Principle of Population*, in which he draws attention to the contrast between life's fecundity ('nature has sowed the seeds of life abroad with the most profuse and liberal hand') and yet how relatively few offspring survive to adulthood, most having perished in the 'struggle for existence'.

The implications, Darwin later recalled, 'struck me at once': it was not just the strongest and most robust that would survive the 'struggle for existence', but those which possessed some special attribute or variation denied their less successful siblings, conferring an advantage by making them more suited to the demands of their environment. They, and their offspring in their turn, were more likely to 'make it', by inheriting that advantageous variation. Repeat that process over several generations, and the advantage would become progressively more refined, and those possessing it better adapted still. 'The result would be the formation of a new species,' Darwin noted. 'Here at last I had a theory [natural selection] by which to work.'

For the next twenty years Darwin would weave the thread of Natural Selection into a web of cogent argument – but held back from publishing a theory that so clearly contradicted the powerful prevailing orthodoxy of Cuvier's 'laws of similarity and correlation' and Paley's 'argument from design'. He was eventually compelled to act in 1858, when his fellow countryman and naturalist Alfred Wallace, laid low with malaria in a Malayan jungle, had precisely the same insight, linking the phenomenon of variation with the struggle for existence as the basis for a scientific theory of evolution. Darwin, fearful that the all-important priority might go to Wallace ('so all my originality, whatever it may amount to, will be smashed'), arranged for their findings to be presented jointly at a meeting of the learned Linnean Society. It would however be Darwin who would receive 'the accolade of history', with the publication the following year of *On the Origin of Species*.

Darwin and Newton Compared

Darwin's genius, like Newton's, was to perceive the hidden significance of an everyday phenomenon so obvious as to have seemed scarcely worth commenting on. There could be nothing more obvious than that apples fall from trees, but not that the force causing them to do so might stretch up to the moon, and so explain the movements of 'heavenly bodies'. Similarly, Darwin inferred from the commonplace of the subtle variations between closely related but distinct species the revolutionary theory that they were not immutable, variations on a theme of a grand designer's blueprint, but rather that 'nature' would select those with advantageous attributes or variations – and so one species might change into another.

> '[It] would be a most extraordinary fact if no variations had occurred useful to each being's welfare,' Darwin wrote; 'assuredly [such] individuals will have the best chance of being preserved in the struggle for life; and will tend to produce offspring similarly characterised. This principle of preservation, or the survival of the fittest, I have called Natural Selection.'

And he was absolutely right. This process of 'speciation' from a common ancestor accounts very well for those finches with their different-shaped beaks. Numerous subsequent observations have confirmed the principle of natural selection, most notably in the Hawaiian archipelago, where the initial colonisation of the islands by single species of snails, moths, beetles and wasps has given rise to large numbers of derived species: seven hundred unique species of fruitfly, twenty-two unique species of honey creepers, and so on.

But Darwin's *Origin* would never have become 'one of the most important books ever written' if he had merely postulated 'natural selection' as an explanation for these minor '*micro*evolutionary' variations. His reputation rests on his having made the further and vastly greater imaginative leap to argue that the same principle is also the cause of '*macro*evolution', linking all of life together into a common 'lineage of descent' all the way back to the beginning. Here mammals – rabbits and kangaroos, apes, elephants, hyenas, whales and ourselves – can all trace their ancestry back (*or are descended by modification from*) a postulated small, shrew-like animal that existed a hundred or so million years ago.

They in turn belong to a class of vertebrates that includes birds, fish and reptiles, which can all trace their lineage back to (*are descended by modification from*) their common ancestor, an unimaginable five hundred million years ago, and so on all the way back to that first primordial form of life three and a half billion years ago. Turn that process on its head, and Darwin's audacity becomes clearer still. Start with some diminutive early reptile, and try to imagine the near-infinite number of variations that would be necessary for it to 'talk its way up', through myriad minor variations in form, to those bizarrely massive dinosaurs such as tyrannosaurus, with its unnecessarily massive twenty-five-foot tail, or the pterodactyl, with a wingspan the size of a small aeroplane. Start again with a shrew-like mammal and try to imagine again the intermediary forms that would be necessary for it to become a mole, kangaroo, flying bat or ocean-navigating whale.

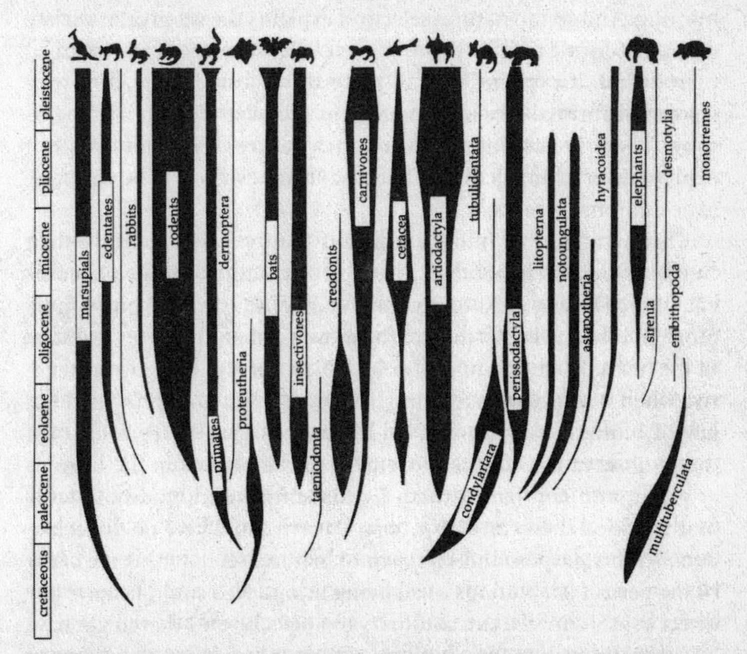

Fig 4-6: *The Mammalian Radiation. Mammals filled the niche left by the extinction of the dinosaurs by diversifying into a plethora of different forms – elephant and giraffe, jumping kangaroo and burrowing mole, rhinoceros and rodent – while primates took to the trees, whales to the sea and bats to the air, and all within the space of twelve million years.*

Darwin may have hesitated before making that great imaginative leap, but its intuitive intellectual appeal was obvious: the extrapolation from 'micro' to 'macro' evolution transforms his proposed mechanism of natural selection from being a theory accounting for minor differences between closely related species into a Master Theory – a universal law that informs every aspect of biology *in its entirety*: the birds in the sky, fish in the sea, mammals on the surface of the earth and 'every creeping thing' beneath it, every type of plant life, trees and grasses and flowers. It explains the perfection of design that so impressed Paley: why eyes are so brilliantly designed for seeing, and ears for hearing. It explains why the parts and organs of those living things are so beautifully adapted to their environment, why the wings of birds and insects are so beautifully adapted for flying, the powerful limbs of predators so well designed for hunting. And so too natural selection explains the yet greater variety of extinct forms of life, the lumbering dinosaur and wide-spanned pterodactyl. It explains how all these things came to be, how they were transformed one into another all the way back to the beginning. There is nothing, in short, that natural selection does not explain. Everything is as it is because it has evolved to be that way over millions of years.

There could be no more ambitious theory, vastly more so than Newton's (relatively humble) theory of gravitation, for the phenomena of life Darwin seeks to account for in *The Origin* are profoundly more diverse and complex than the movement of the moon and stars in the heavens. The comparison with Newton is also highly instructive when it comes to evaluating the evidence for Darwin's universal law of biology. Newton derived his laws of gravity from his own direct observations of the movement of the planets in the heavens at night, and confirmed them by his simple and readily testable mathematical theorem. By contrast, Darwin could have no direct evidence of his proposed mechanism of 'natural selection' as the cause of the perfect adaptations of all living things. Nor could he have any direct evidence of the extraordinary biological events that might have one species transforming itself into another hundreds of millions of years in the distant past. He could only suggest 'imaginary illustrations' such as how, for example, a land-based bear might become 'almost like a whale':

> In North America the black bear was seen swimming for hours with widely open mouth, thus catching, like a whale, insects in the water. Even in so extreme a case as this, if the supply of insects were constant, and if better adapted competitors did not already exist, I can see no difficulty in a race of bears being rendered, by natural selection, more and more aquatic in their structure and habits, till a creature was produced as monstrous as a whale.

Again, in Newton's world, proportional causes generally have proportional effects; so, by definition, a greater force is required to give an object a greater acceleration. In Darwin's world, by contrast, the same cause has grossly disproportionate effects. The same evolutionary process that might explain such relatively trivial differences as the shape of a finch's beak becomes responsible for the prodigious differences that separate a mouse from an elephant, an octopus from a bee. Here the great virtue of Darwin's proposed mechanism, its simplicity, might seem its greatest drawback – that it is *far too simple* to begin to account for the complexities of life. His theory puts the details of those complexities to one side, subsuming them within a *historical* explanation of how the living world has come to be as it is.

Further, his audacity in proposing so all-encompassing a theory presupposes two biological processes that might reasonably be thought (and certainly were at the time) to be impossible. Paley's argument of the necessity for a Grand Designer may not be convincing, but it is indeed a puzzle how and why the eye's many distinct parts should each be so perfectly designed for seeing. Here the *least likely* of possible explanations had to be that that 'perfection' was the consequence of the random accumulation of numerous small, fortuitous variations. Similarly, Darwin's supposition of a gradual, scarcely detectable, transformation from one form of life to another would appear to contradict the entire body of knowledge of natural history at the time, and particularly Cuvier's law of correlation, which held that no *part* can change without the *whole* changing.

> 'It is possible,' Cuvier had observed, 'to determine the class of an animal after inspecting a *single* bone . . . this is because the number, direction and shape of the bones that compose each part of an animal's body are always in a necessary relation to all the other parts, in such a way that one can infer the whole from any one of them.'

The significance of these two 'sticking points' – the puzzle of perfection and gradualist transformation – is reflected in their being described as axiomatic. Darwin's theory, which holds that they are the consequence of the accumulation of numerous small beneficial variations, must account for both. If it does not, his theory falls, and it becomes necessary again to presuppose some powerful but unknown 'organisational' principle, similar to Cuvier's laws, orchestrating the complexities of life. These two axioms would, as will be seen, prove very troublesome.

A Sceptical View

It is not immediately obvious how Darwin could successfully have pulled off so staggering an intellectual leap on such slender evidential grounds. The prevailing view would be that he succeeded because he was right, or as right as the knowledge of his time would permit – and that all subsequent scientific findings have tended to confirm the principle of natural selection as the true cause of the diversity of life.

There is, however, more than enough evidence already to suspect that Darwin was less right than is commonly perceived, while the subsequent success of his theory fails to reflect the well-grounded scepticism of many of his scientific contemporaries. This was, after all, the Golden Age of Natural History, to whose cornucopia of discoveries Darwin's contemporaries had made their own unique contributions, describing numerous new species of plants and animals and reconstructing from fossilised bones and teeth the form of extraordinary creatures long since extinct. That generation had, in short, a feeling for the complexities of biology, how things fitted together, which defied any simple explanation – and none more so than 'the British Cuvier', Richard Owen, founder in 1881 of the largest natural history museum in the world, author of over 360 papers on fossil species, and the first person to coin the term 'dinosaur'. 'There was not,' observed *The Times*, 'a more distinguished man of science in the country.' His opinions were therefore likely to be particularly well informed.

Darwin's central thesis, Richard Owen argued in a critical review of *The Origin*, was 'no common discovery, no very expected conclusion

[to]...link together worms and fleas, moles and elephants, turnips and men as all lineal descendants of the same [unknown] common ancestor'. He acknowledged that nature does, of course, 'select': it eliminates the feeble and the weak that are unable to fend for themselves, and thus promotes the health and robustness of all forms of life. This might, at the simplest practical level, account for why antelopes run so fast, the better to escape their predators. It was, however, a different matter to suppose that the same mechanism that weeded out the unfit could also be the creative force, as Darwin claimed, the *sculptor* of the infinite diversity of form and attributes of creatures living and extinct: 'Natural selection is daily and hourly scrutinising every variation, even the slightest, rejecting that which is bad, preserving and adding up all that is good; silently and insensibly working at the improvement of each organic being.'

But, Owen asked in the first of a series of perceptive criticisms, what could be the advantage in nature selecting the 'simplest variation'? There would need to be 'numerous and successive' such variations to further a species' prospects for survival. What then are the chances that one simple variation would be followed by another similar one to reinforce it, and another, and another, in such a way as to change or transform one species into another? Nor is it sufficient for nature to 'select' those small variations; parents must also pass them on to their offspring – which is certainly not the case, for that would require children to be their parents' exact replicas. And even if nature were to select these numerous slight variations, and they were passed on to the next generation, sooner or later they would encounter the barrier of 'hybrid sterility', which prevents the interbreeding of one species with another – ensuring that dogs stay dogs, and pigeons pigeons, and their progeny also.

Darwin might suppose this to occur, observed Owen, but:

> In the name of all true philosophy we protest against such a mode of dealing with nature. We are not offended in the 'Arabian Nights' at an impossibility, when Amina sprinkles her husband with water and transforms him into a dog – but we cannot open the doors of the temple of scientific truth to the genii and magicians of romance. [It is the] true spirit of philosophy to be reluctant to admit to great changes suggested by the imagination – but the steps of which we cannot see.

There was not, in short, the slightest evidence for Darwin's 'not very expected' conclusion – on the contrary:

> We have searched in vain for the evidence that it is only necessary for one individual to vary, be it ever so little, in order to [validate] the conclusion that variability is progressive and unlimited, so as, over the course of generations, to change the species, the genus, the order or the class. We have no objections to 'natural selection' in the abstract; but we have desire to have reason for our faith. What we do object to is that the true character of science should be compromised by *mere hypothesis.*

But Darwin's 'mere hypothesis', as we know, won the day, and triumphantly so. And for why? First, he presents his argument in *The Origin* in such a way as to suggest that the evidence in its favour is much more compelling than was in fact the case. And, second, the timing was perfect. *The Origin* articulated the desire of many scientists for an exclusively materialist explanation of natural history that would liberate it from the sticky fingers of the theological inference that the beauty and wonder of the natural world was direct evidence for 'A Designer'.

THE TRIUMPH OF THE ORIGIN*: PART 1*

Ingenuity of the Argument

On the Origin of Species opens with Darwin, the youthful naturalist, struck by the profound insight which had eluded his eminent contemporaries – that 'species were not immutable'.

> When on board HMS *Beagle* I was much struck with the distribution of the inhabitants [species] of South America . . . that seemed to me to throw some light on the origin of species – that mystery of mysteries . . . I am fully convinced that species are not immutable; but are lineal descendants of some other and generally extinct species . . . The view that most naturalists entertain and which I formerly entertained, namely that each species has been independently created – is erroneous.

The resolution of that 'mystery of mysteries' requires first that there be a wide range of variation within any individual species, from which nature can select the most biologically advantageous – for which, Darwin points out, there is abundant evidence in the popular nineteenth-century hobby of competitive pigeon-breeding. Over a couple of hundred years pigeon fanciers had successfully transformed the plain-looking standard rock pigeon into an extraordinary medley of different forms: the fantail, the Jacobin, the trumpeter, the pouter (with an enormously developed chest which it glories in inflating), the short-faced tumbler and the carrier. Why, asked Darwin, if man can breed from a single stock so many different forms, should not 'nature', by selecting the most advantageous of variations, produce species 'so infinitely better adapted to the most complex conditions of life, that plainly bear the stamp of far higher workmanship'?

His own observations from the Galapagos Islands suggested just such 'microevolutionary', nature-induced adaptations as an explanation for those diverse species of finches. The challenge is to extrapolate that same principle to the whole of biology. Here, Darwin's ingenuity lies in applying two quite different forms of logical argument, first by *minimising* the difficulties posed by 'macroevolution', and particularly those two axiomatic principles that natural selection would have to account for – the 'puzzle of perfection' and the practicalities of gradualist transformation. He then *maximises* the significance of those microevolutionary changes by contrasting the much greater plausibility of his explanation with that in the version of Genesis, where God brings into existence 'every living thing that moveth' on the fourth and fifth days of His creation of the world. This will become clear enough.

The task of minimising the difficulty posed by the first axiom – natural selection as an explanation for the 'puzzle of perfection' – could not be more formidable. The problem, as illustrated by the eye, is twofold. First, as Paley had pointed out, each separate component part (lids, cornea, iris, lens, retina, etc.) has to be exquisitely designed for its specialist purpose, for if any is slightly less than perfect, then, like a defective cog in his watch, the eye would (as it were) be unable to tell the time. Yet each of those exquisitely designed specialist parts by itself is useless, and only finds 'meaning' within the context of the eye in its entirety. The same applies to the infinite variety of the perfectly adapted distinctive features of all living things, each composed of numerous components designed for a specialist purpose which can only find 'meaning' within the context of the whole – wings for flying, ears for hearing, hearts for pumping, lungs for breathing, and so on *ad infinitum*.

Further, it is necessary to emphasise the *extreme* perfection of the eye. It is not merely 'well designed' – its sensitivity is so acute that it cannot be improved upon: capable of detecting the faint light of a night fire at a distance of several miles, and accommodating levels of brightness from sunlight on ice to flickering shadows by night. By comparison, Paley's watch is a humble thing indeed.

How, then, can nature achieve what in everyday human experience would be thought a miracle – a watch designed to the highest possible specifications – without a designer? Darwin concedes the problem:

> 'Nothing at first can appear more difficult to believe than that the more complex organs have been perfected not by means analogous with human reason [i.e. having been designed for their purpose], but the accumulation of innumerable slight variations each good for the individual possessor [i.e. each separately conferring some biological advantage].'

There is, he acknowledges, nothing more difficult to believe than that the eye should be merely a consequence of such a random process:

> 'To suppose that the eye, with all its inimitable contrivances for adjusting the focus [the lens], for admitting different amounts of light [the iris]...could have been formed by natural selection, seems, I freely confess, absurd in the highest possible degree.' And not just absurd: 'My theory would break down . . . if it could be demonstrated that any complex organ could not have been formed by numerous slight modifications.'

But Darwin 'can find no such cases'. Indeed, he has 'no very great difficulty' in supposing that natural selection might have given rise to the 'perfect optical instrument' of the eye – which, moving from the simplest form of a light-sensitive spot, would then, 'graduating in diversity', become ever more complex, adding first the lens, then a shutter-like iris, and so on, culminating in the human eye.

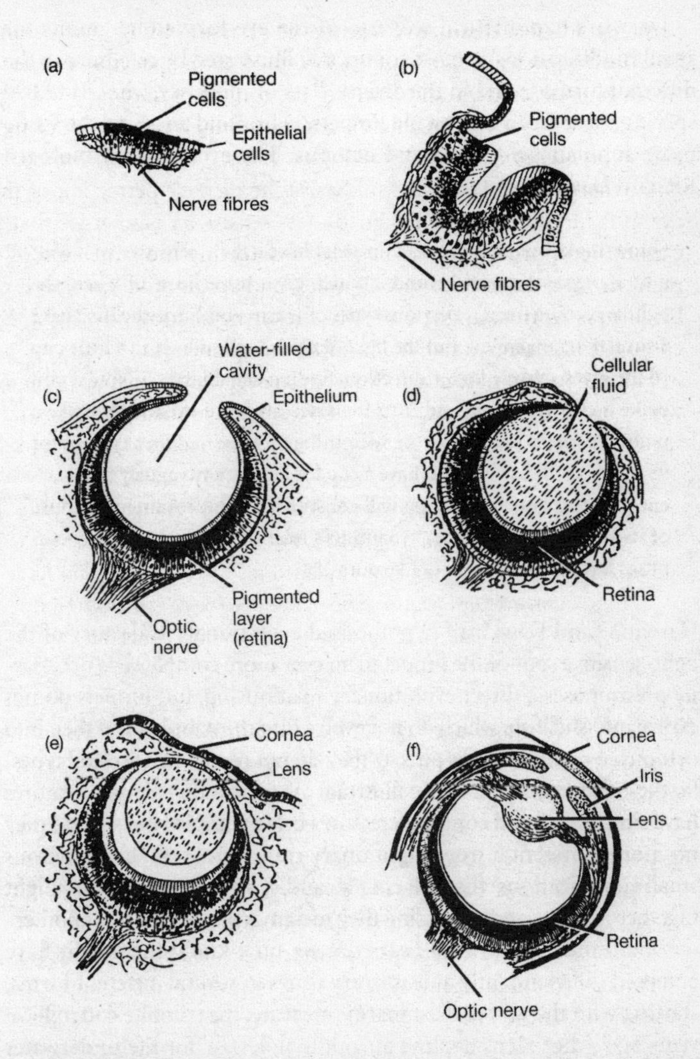

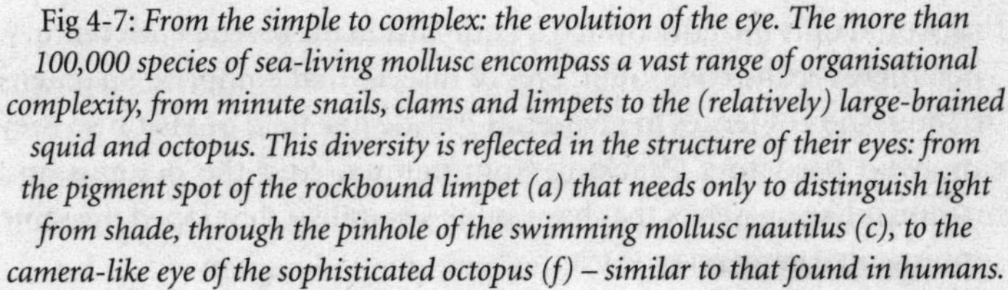

Fig 4-7: *From the simple to complex: the evolution of the eye. The more than 100,000 species of sea-living mollusc encompass a vast range of organisational complexity, from minute snails, clams and limpets to the (relatively) large-brained squid and octopus. This diversity is reflected in the structure of their eyes: from the pigment spot of the rockbound limpet (a) that needs only to distinguish light from shade, through the pinhole of the swimming mollusc nautilus (c), to the camera-like eye of the sophisticated octopus (f) – similar to that found in humans.*

Darwin's hypothetical scenario of the eye formed by 'numerous small modifications' is most commonly illustrated by reference to the different forms found in the diverse class of molluscs, whose 112,000 species range from minuscule limpets, clams and snails to the vastly more sophisticated squid and octopus. The evolutionary biologist Richard Dawkins elaborates:

> Some single-cell animals [and limpets] have a light-sensitive spot with a little pigment screen behind it which gives some 'idea' of where the light is coming from . . . various types of worm and some shellfish have a similar arrangement, but the light-sensitive cells are set in a little cup [that] gives slightly better direction-finding capability . . . Now if you make a cup very deep and turn the sides over you eventually make a pinhole camera [such as in the swimming mollusc nautilus, famous for its twisty shell]...when you have a cup for an eye, any vaguely transparent material over its opening will constitute an improvement because of its lens-like properties. . . [nautilus's relatives] the squids and octopuses have a true lens, very like ours . . .

Darwin's (and Dawkins') hypothesised evolutionary trajectory of the light-sensitive spot of the limpet to an ever more complex eye necessarily presupposes a direct evolutionary relationship, but limpets do not evolve into shellfish, which do not evolve into the nautilus and then into octopuses; or, as Dawkins puts it, 'they do not represent ancestral types'. So these different types of eye illustrate the truism that simple creatures have simple eyes and complex creatures more complex ones – and they no more represent a true evolutionary transformation 'by numerous small modifications' than placing a candle, a torch and a searchlight together in a row and supposing they metamorphose into one another.

One hundred and fifty years on, we now know the eye to have emerged independently at least forty times in several different forms, starting with the very earliest marine creature, the trilobite 450 million years ago, whose lens deploys an 'optimal design' for life underwater that would only be rediscovered by humans in the seventeenth century. Then there are the compound eyes of insects, that comprise up to tens of thousands of lenses to give a full 360-degree field of vision so they can avoid predators attacking from behind. And the octopus and mammals like ourselves that have, quite separately, developed the same sophisticated camera-type eye.

Each different type of eye compounds Darwin's difficulty further, for then it is necessary to presuppose for each a series of fortuitous 'numerous successive slight modifications', conferring some slight biological advantage to its possessor. It is necessary to *presuppose*, for, despite much effort, there is not a single empirical discovery in the past 150 years that has substantiated Darwin's proposal that natural selection, 'taking advantage of slight successive variations', explains the 'puzzle of perfection' epitomised by so many different types of eye – which remains yet more puzzling than it was in 1859.

The difficulty posed by the second axiom, the practicalities of how natural selection brings about the gradualist transformation of one species into another, is precisely the reverse. Now there is an abundance of empirical fact in the fossil record to test the theory's validity. That process of gradualist transformation requires, as suggested, there to be a near-infinite number of intermediary species 'talking their way up' from the first single-cell organisms to the near-infinite variety of forms of dinosaur or mammal.

> 'The number of intermediate and transitional links between all living and extinct species, must have been *inconceivably* great,' Darwin writes, 'and assuredly, if [my] theory be true, such have lived upon the earth.'

The problem is that the history of life, as told in the fossil record, reveals the contrary pattern of the sudden emergence (in successive waves) of a diversity of new life forms; their persistence, virtually unchanged, over millions of years; and then their sudden and unexplained disappearance – only for the whole cycle of emergence, stability and extinction to be repeated. Thus, a brief synopsis of the 'History of Life' starts three thousand million years ago with those first single-cell organisms. Then nothing much happens till six hundred million years ago with the 'Cambrian explosion' of marine fossils, which culminated in their mass extinction 250 million years later. This was followed by the 'dinosaur explosion' that lasted till *their* mass extinction seventy million years ago. And this in turn was followed by the 'mammalian explosion' of which we are a part. Meanwhile, along the way the major groups that mark the crucial

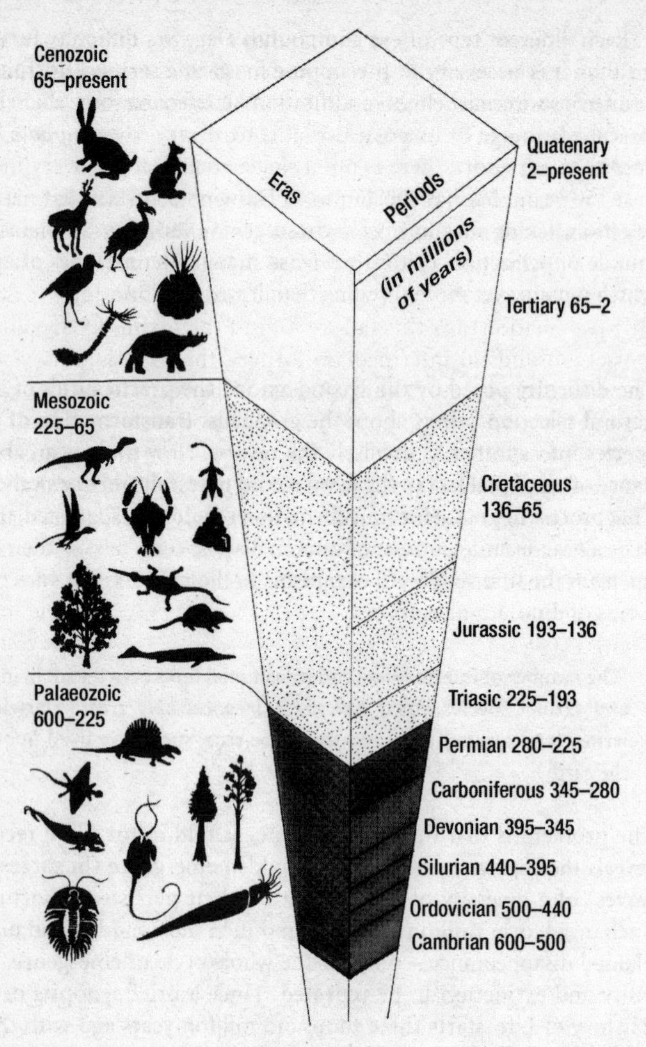

Fig 4-8: *The history of life divides into three 'eras' – running up the left-hand side of the column – that correspond with the three main 'radiations' of forms of life, starting with the Palaeozoic (ancient) radiation of marine creatures at the beginning of the Cambrian period – known as the 'Cambrian explosion'. This was succeeded by the Mesozoic (middle) radiation of the dinosaurs, that were in their turn supplanted following their extinction by the Cenozoic (recent) mammalian radiation of sixty-five million years ago.*

transitions from sea to land and land to air make their appearance with little or no warning. 'It is as though life goes behind the bushes and emerges in new clothes,' writes the biologist Robert Wesson of Harvard University – and in an abundance of diversity that defies all imagination.

Darwin concedes that 'all our most eminent palaeontologists' (and he cites half a dozen) and 'our greatest geologists' were unanimously opposed to the notion of gradualist transformation for just this reason. Nor were they convinced by his explanation (and it is the only explanation) for this failure to find the fossilised remains of those thousands of intermediary forms – that they had not as yet been discovered due to the 'extreme imperfections of the fossil record':

> I look at the geological record as a history of the world imperfectly kept and written in a changing dialect; of this history we possess the last volume alone relating only to two or three countries. Of this volume only here and there a short chapter has been preserved; and of each page only here and there a few lines. Each word . . . may represent the forms of life.

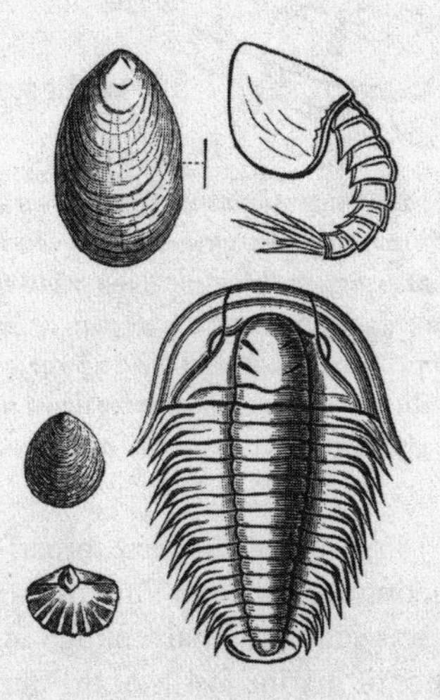

Fig 4-9: *The many diverse forms of life of the Cambrian explosion included the trilobite, the free-swimming oceanic winged snail, the shrimp-like hymenocaris and a medley of clam-like creatures.*

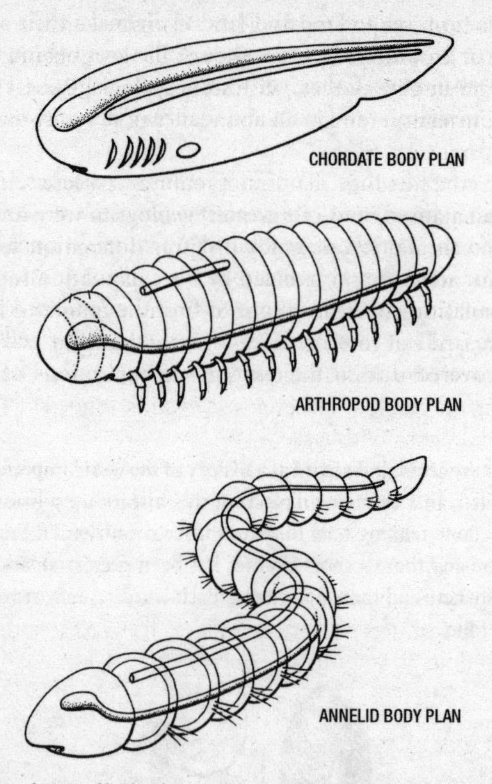

Fig 4-10: *A profusion of body plans. All forms of life are 'variations on a theme' of several distinct body architectural plans, virtually all of which were established by the close of the Cambrian explosion (life's 'Big Bang'). The three illustrated here are, from the bottom upwards, the segmented body plan as seen in the earthworm; the arthropods such as insects and crabs with their typical jointed limbs; and the chordate vertebrates with a spinal column running down the back that would in time give rise to fish, mammals and humans.*

The fossil record in 1859 was indeed very imperfect, but as one of Britain's leading palaeontologists, John Phillips, would point out the following year, there was still more than enough in the 'first wave' of life, the rich abundance of marine fossils of the Cambrian explosion, to test Darwin's hypothesis. The findings were quite unambiguous. The strata of rocks are empty, and then suddenly, from apparently nowhere at all, they are filled with the fossilised remains of billions upon billions of complex forms of life – clams, sea urchins, crustaceans with jointed legs, and most impressive of all, the resilient trilobite that comes complete with eyes, and distinct head and tail, and sticks around virtually unchanged for 250 million years before vanishing as inexplicably as it arrived.

Further, the fossils of that first 'Cambrian explosion' include not just representatives of the three main *body plans* of living species – worms, insects and vertebrates, with quite different patterns of segmentation, nervous and circulatory systems – but also a phantasmagoric collection of creatures that defy any attempt to place them in any recognised category, including the aptly named *hallucigenia*, so called because of its 'bizarre and dream-like appearance', that propelled itself along the sea floor on seven sets of stilts, echoed by seven tentacles protruding from its back.

Still, Darwin was right that there was no telling what the future might reveal, and *The Origin* proved to be the greatest boon for future generations of geologists and palaeontologists, by focusing their (and the public's) attention on the search for those intermediaries that would reveal the all-important transitions between the major categories of life from those marine fossils to fish, fish to amphibians, to reptiles, and finally to birds and mammals.

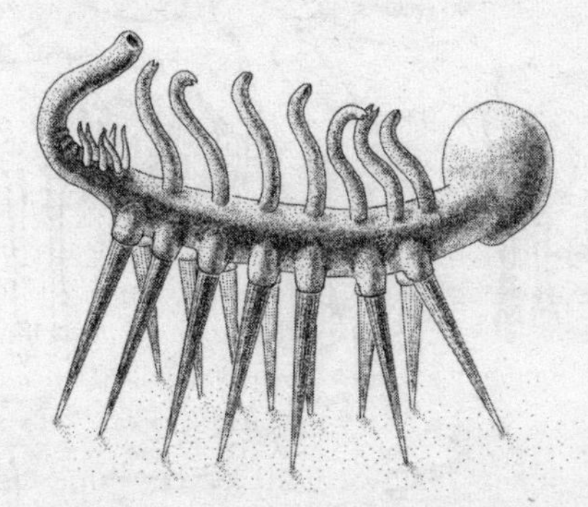

Fig 4-11: *The Burgess Shale, a quarry high in the Canadian Rockies, contains the most complete fossilised record of the 'Cambrian explosion', with more varieties of life than can be found in our modern oceans. These include a plethora of extraordinary forms – such as* hallucigenia, *illustrated here, and* opabinia, *with five eyes on stalks, a backward-facing mouth and a long, flexible, hose-like protuberance that ended in a claw fringed with spines.*

Almost immediately, it seemed as if Darwin had confounded the 'unanimous' opinion of those eminent palaeontologists when the following year geologists in Bavaria discovered the remarkable *archaeopteryx*, which appeared to mark the transition from reptiles to birds, with reptilian features of teeth, a tail and claws at the end of its feathered wings. Fourteen years later, geologists in the United States would link together the bones and teeth of fossilised horses to produce the most familiar instance of Darwin's gradualist transformation, starting with the fox-sized *eohippus*, which over fifty million years gradually increased in size, while its feet underwent the transition from five toes to one, so characteristic of the modern horse. Most compelling of all would be the evidence for the Ascent of Man, where the discovery of the beetle-browed 'caveman fossil' Neanderthal man in the Neander Valley in 1856, when placed in sequence with the skeletal remains of the first truly modern human, Cromagnon man (discovered in 1868), provided that seemingly coherent picture of man's evolutionary trajectory.

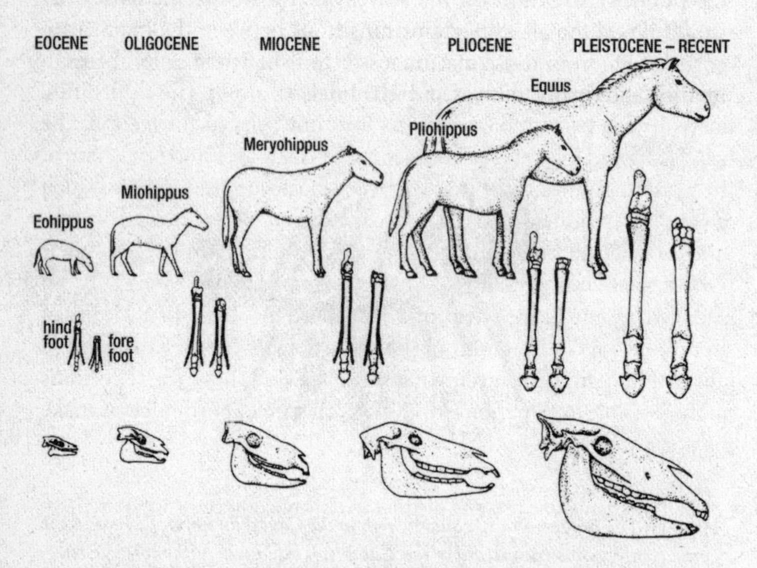

Fig 4-12: *From* eohippus *to* equus. *In the mid-1870s the American palaeontologist Othniel Charles Marsh discovered a series of fossils delineating the upward progression in size from* eohippus *('dawn horse') to the modern horse,* equus.

These examples are familiar enough, but they are frequently cited for the very good reason that there are so *few* instances where the evolutionary history of a species is suggestive of gradualist transformation. The fossil record, over the next 150 years, would become ever more complete (and thus less 'imperfect'), but that 'inconceivably great number of transitional species' that might mark the transformation of a small reptile into a dinosaur, or a shrew-like mammal into a kangaroo, remained as elusive as ever. Instead, the fossil 'story' reads much as those eminent nineteenth-century palaeontologists such as John Phillips had interpreted it:

> 'Most people assume that fossils provide important [evidence] in favour of the Darwinian interpretation of the history of life. Unfortunately, this is not strictly true,' observes David Raup, of the Natural History Museum in Chicago. 'Rather than the gradual unfolding of life . . . species appear in the sequence very suddenly, show little or no change during their existence in the record, then abruptly [disappear].'

* * *

Darwin changes tack in the second half of *The Origin*. The 'difficulties' posed by his two fundamental axioms now placed to one side, he brings forward his observations from the Galapagos and elsewhere of the relatively minor differences between closely related species – such as the finches' beaks – and contrasts his reasonable natural 'microevolutionary' explanation with the Biblical supposition in Genesis of their having been individually created by God. This is essentially an extended version of the scientific (and indeed theological) absurdity of supposing that God had nothing better to do with his time than design ten thousand different species of beetle. Darwin's rhetorical reflections on the possible explanations for the several species of rhinoceros make the point well enough:

> Shall we then allow that the distinct species of rhinoceros which separately inhabit Java and Sumatra were created, male and female, out of the inorganic materials of those countries? Shall we . . . say that without any apparent cause they were created on the same generic type with the ancient woolly rhinoceros of Siberia and all the other species which formerly inhabited that part of the world. . . [shall we say] that without

> any apparent adequate cause their legs should be built on the same plan as those of the antelope, of the mouse, of the monkey . . . I repeat shall we then say that a pair of each of these three species of rhinoceros were separately created with the deceptive appearance of a true relationship.

Darwin, over the years, had amassed much evidence in favour of microevolution as the cause of these minor variations on a theme of individual species. Time and again he returns to the same point:

> We see that nearly all the plants and animals of the Galapagos archipelago . . . being related in the most striking manner to the plants and animals of the neighbouring American mainland. *It must be admitted these facts receive no explanation on the theory of creation.*
>
> We can see why oceanic islands should be inhabited by few species . . . and that [new and peculiar] species of bats, which can traverse the ocean, should so often be found on islands far distant from any continent. *Such facts are utterly inexplicable on the theory of independent acts of creation.*
>
> The inhabitants of the Cape de Verde islands are related to those of Africa, like those of the Galapagos to America. *This grand fact can receive no sort of explanation on the ordinary view of independent creation.*

And, referring to the common plan or 'homology' of the limbs of diverse species as described by Baron Georges Cuvier, Darwin argued:

> What can be more curious than that the hand of man, formed for grasping, that of a mole for digging, the paddle of the porpoise and the wing of the bat should all be constructed on the same pattern, and should include similar bones, the same relative positions. *Nothing can be more hopeless than to explain this similarity of pattern [by supposing] it has pleased the Creator to construct all the animals in each great class on a uniform plan.*

Here Darwin, in portraying those who might dispute his explanations as being Biblical creationists, is clearly setting up a false antithesis. Few biologists could dispute, once it was pointed out, that there must be a natural microevolutionary explanation for the many variations on the theme of closely related species. Cuvier's laws of similarity and correlation, with all their implications for the much larger question of the mechanisms of macroevolutionary transformation, were clearly a different matter – and could scarcely be described as 'creationist', with all its fundamentalist connotations. Rather, Cuvier's laws were grounded in the science of comparative anatomy, and put to the test by his reconstruction of that megatherium from the bones and teeth of its fossilised remains. To be sure, he inferred a hidden metaphysical 'idea' behind that blueprint, but Darwin's implication that this undermined his scientific credibility while legitimising his own much simpler materialist explanation was clearly unwarranted.

The Origin is long and at times heavy going, and many readers would no doubt have been tempted to skip along, and so would not have noticed when reaching the Conclusion that Darwin's argument had undergone a further metamorphosis, to become yet clearer still. Now the theory of natural selection accounts for 'most of the leading facts' in natural history, while the 'difficulties' are so finessed as no longer to be difficulties at all. Though the fossil record is still 'imperfect to an extreme degree', nevertheless, the reader learns, 'such that the record gives' *supports* the theory of descent with modification. There is even a walk-on part for the Creator, to initiate the whole process by imprinting the laws of natural selection onto nature – thus giving a theological gloss to Darwin's triumphant and frequently cited final paragraph:

> There is grandeur in this view of life, with its several powers, having been originally breathed into a few forms or into one; and that whilst this planet has gone on cycling according to the fixed laws of gravity, from so simple a beginning endless forms most beautiful and most wonderful have been evolved.

And that, in a nutshell, is how Darwin pulled off the greatest intellectual coup in science, fashioning an all-encompassing theory of gradualist evolution to resolve the 'mystery of mysteries' in defiance of all the theoretical considerations and empirical evidence that would contradict it. But the timing was perfect.

THE TRIUMPH OF THE ORIGIN: *PART 2*

The World was Ready for the Theory of Evolution

'The immediate success of *The Origin* indicated the world was ready for the Theory of Evolution,' writes veteran biologist Ernst Mayr. Britain presided over a vast empire covering a quarter of the globe, many of whose citizens could no longer be doing with the God of Genesis creating the world in six days, who then (the fossil record suggested) went on creating ever more waves of new species at intervals of hundreds of millions of years, before causing them to become extinct and starting all over again. They were more than ready for Darwin's 'better' theory, whose precise details hardly mattered. Who were non-scientists to judge, one way or another, the validity of his explanation? It seemed plausible, and if there were gaps in the fossil record, no doubt the forward march of science would fill them in due course.

Most of Darwin's scientific contemporaries, as suggested, were less easily persuaded. But these were scientifically exhilarating times, when the discovery in the previous fifty years of the foundations of both Chemistry (epitomised by Lavoisier's finding that water is composed of the two chemical elements of hydrogen and oxygen) and Physics (the nature of heat, light and energy) had established both as independent, autonomous scientific disciplines. Within this context, the continuing influence of theological ideas on Natural History was increasingly anomalous. Further, for all Richard Owen's persuasive contrary arguments, it made little difference to most biologists in their everyday lives whether Darwin was right or not. The imperative to believe in the *principle* of evolution by natural law more than outweighed its obvious deficiencies: 'We accept [the theory of natural selection] not because we are able to demonstrate the process in detail, nor indeed because we can with more or less ease imagine it,' observed his contemporary the zoologist August Weismann, 'but simply because we *must*, because it is the only possible explanation that we can conceive.'

> 'Thus we find ourselves in a singular position,' observed the French palaeontologist François Jules Pictet in 1860. 'We are presented with a theory which on the one hand seems to be impossible because it is

> inconsistent with the observed facts. But, on the other hand, it appears to be the best explanation of how organised beings have developed in the epochs previous to ours.'

Within a decade the majority of biologists had endorsed at least the principle of a materialist evolutionary explanation along Darwinian lines, bolstered by the findings of those transitional forms such as *archaeopteryx* that his theory had predicted, and which though not 'accessible to proof', were 'yet in general harmony with scientific thought'. And so the campaign progressed to 'wrest from theology all schemes and systems that infringe upon the domain of science'.

The mark of any truly powerful and influential idea is its ability to be 'all things to all men', and contrary to the conventional view, the Church and its leading spokesmen found much in the evolutionary theory to theology's advantage. It offered a way out from the continuing embarrassment of having to endorse a literal belief in the Biblical version of the creation story, while the proposal that God might exert his influence through the evolutionary mechanism seemed a more sophisticated, and certainly more acceptable, version of Paley's 'argument from design'. Indeed, the evolutionary theory could be interpreted as an active inducement to religious belief by resolving that great stumbling block to faith: how an all-powerful (and all-loving) God should tolerate the cruelty, evil and unhappiness in the natural world and human affairs. Nature is replete with examples of seemingly gratuitous and inexplicable cruelty – such as a species of predator ant whose limbs have a serrated edge with which it literally saws off the head of its prey. Some might suppose this 'problem of evil' to be a more powerful argument against belief in a beneficent Creator than the beauty and wonder of the world being evidence in His favour. And now Darwin offered at a stroke to resolve this paradox by absolving God of direct responsibility for evil in a divinely created world. God could scarcely be expected, having impressed the laws of natural selection on life, to then miraculously intercede to forestall or prevent any misfortunes that might happen as a result. 'Darwin appeared under the guise of a foe but did the work of a friend,' remarked the liberal Anglo-Catholic Aubrey Moore. 'He conferred upon philosophy and religion an inestimable benefit.'

* * *

The appeal of *The Origin* to so wide a spectrum of opinion of the public, scientists and even the Church might seem reason enough for its surprisingly rapid acceptance. But this scarcely accounts for why Darwin's speculative theory should become the foundational theory not just of biology, but of science itself. For that it is necessary to dig deeper, and to recognise that the significance of *The Origin* lies far beyond the realms of biology, as the lightning rod for a whole set of political and philosophical ideas generally subsumed under the heading of 'the Enlightenment'.

The roots of the Enlightenment lie in the mid-seventeenth century, when, in reaction to the vindictive wars between Catholic and Protestant that had brought religion into disrepute, the scientific revolution of Galileo and Newton held out the promise that human reason, rather than divine revelation, offered the most reliable guide to knowledge and future progress. The Enlightenment's optimistic confidence in the individual and collective power of ordinary people to change their lives for the better seemed much preferable to that religious fatalism which holds human life to be no more than a staging post on the way to fulfilment in the next world.

Those seeds of Enlightenment optimism were in turn fed and watered by the French philosophers Diderot and Voltaire, who, challenging the authority of Church and State, argued for the principles of human self-determination epitomised by the victory of the fledgling United States in their War of Independence and the overthrow of the absolute monarchy in France by the French Revolution, with its slogan of *Liberté, Égalité, Fraternité*. Meanwhile, the same Enlightenment ideals would find favour in a slightly more nuanced form in Britain and the United States, amongst both the burgeoning middle class seeking to broaden its political influence over the still powerful aristocracy, and the industrial working class struggling to liberate itself from the bondage of wage slavery.

The Enlightenment cut many ways, but by the mid-nineteenth century it had crystallised around three central themes, each of which could be interpreted as being antagonistic to the enduring influence of Christian thought and belief. They were: first, the assertion of human reason as the sole source of reliable knowledge, as against the Biblical revelation of the Word of God; second, nature was perceived as a closed system of causes and effects governed by its own internal laws that were immune to divine intervention – thus precluding the possibility of miracles, whether 'natural' or otherwise; and third, a belief in scientific progress as giving a more direct purpose and meaning to life than the mysticism of religion.

The Origin, suffused by these ideas, was guaranteed a sympathetic hearing. But its influence went much further in crystallising the ascendancy of science, by cutting the knot that still tethered people to the religious intimation, that comes from contemplating the beauty and diversity of the natural world, that there might be 'more' than man can know. Darwin's explanation was in its own way profoundly 'metaphysical', in that it attributed to natural selection powers that might reasonably be thought to be miraculous – that it should somehow fashion perfection from a blind, random process, and transform one class of animal into another without any empirical evidence of having done so. But, crucially, his materialist (if in reality miraculous) explanation belonged to the scientific view that held, as the historian of science William Provine points out,

> the world to be organised strictly in accordance with mechanistic principles, where . . . there are no purposive principles whatsoever in nature. There are no gods and no designing forces that are rationally detectable . . . there are no inherent moral, ethical laws, no absolute guiding principles for human society . . . no hope of life everlasting . . . free will as it is traditionally conceived does not exist . . . there is no ultimate meaning for humans.

Darwin's *Origin* did not just liberate natural history from the sticky fingers of theology, but man from the superstitious beliefs of the past, opening the way to a glorious future of scientific progress inspired by reason, and the promise of human self-determination. The next hundred years would witness an inverse (and causative) relationship in the relative fortunes of science and religion, the rise in the former as the standard-bearer of reason and progress being in direct proportion to the decline of the latter, whose authority and self-confidence were deeply undermined.

This transition from a predominantly religious to a secular worldview has proved almost surprisingly painless, as it turns out that it is not in the least necessary to invoke (or presuppose) some 'higher power' in order to lead a good, exemplary, moral life. But that trumping of religion by science was not without its costs. It created a historical rupture in the continuity of Western civilisation, erecting a glass window separating its present and its future from the wisdom and achievements of its past. 'Prior to 1859 all attempts to answer the question "What is man?" are worthless,' observed the evolutionary biologist Gaylord Simpson. 'We would be better off if we ignore them completely.'

Further, there are dangers in supposing the wonder and diversity of the natural world to be so much more readily explicable than it really is. The greatest obstacle to scientific progress, after all, is not ignorance, but the illusion of knowledge. Darwin's evolutionary theory readily short-circuited serious intellectual enquiry as it could, in an instant, produce a reason for (literally) *everything*. There was nothing it could not explain, not even the paradoxically non-sloth-like toilet habits of the sloth:

> The sloth, instead of defecating on demand like other tree dwellers, saves its faeces for a week or more, not easy for an eater of coarse vegetable material. It descends to the ground it otherwise never touches, relieves itself, and buries the mass. The evolutionary advantage of going to this trouble, involving no little danger, is supposedly to fertilise the home tree. That is [the argument runs] a series of random mutations led an ancestral sloth to engage in un-sloth like behaviour for toilet purposes. This then so improved the quality of foliage of its favourite tree as to cause it to have more numerous descendants than sloths that simply let their dung fall – and thus the trait prevailed.

The Origin rapidly became a foundational text for science not just by having a (materialist) reason for everything, but because it was in essence immune to criticism. Science as the victorious party had the privilege of writing the historical account in its own favour, as one of 'reason and light' triumphing over the superstitious beliefs of the past. By such an account those, like Richard Owen, who challenged the explicative power of natural selection, were readily portrayed as being motivated by their partisan (and thus necessarily biased) prior commitment to the philosophic view, and thus their searching criticisms could be set to one side. The yet more effective defence against such criticism was to concede that *The Origin* was indeed flawed, but nonetheless, as Thomas Huxley insisted, it remained 'the best explanation that had yet been offered – *without claiming that any part has yet been confirmed*'. Darwin's friend, the botanist and director of the Royal

Botanical Gardens at Kew, Joseph Hooker, went further, confessing that he was 'holding himself ready to lay it down should a better explanation be forthcoming'.

This notion that a faulty theory is better than no theory at all should, by rights, have no place in science. 'We live in a small bright oasis of knowledge surrounded on all sides by a vast unexplored region of impenetrable mystery,' the three-time Prime Minister the Marquess of Salisbury would observe when addressing the British Association in 1894. '[But] we are under no obligation to find a theory if the facts will not provide a sound one. To the riddles which nature propounds, the profession of ignorance must constantly be our only reasonable answer.' But that 'profession of ignorance' was scarcely an option for those bent on liberating natural history from theology, for whom a 'materialist explanation', *any* materialist explanation, would do.

We turn now to consider what happened next, but it is perhaps appropriate to recall, in the robust prose of evolutionary biologist Stephen Jay Gould, the scale and the challenge *The Origin* posed to the traditions and beliefs of the preceding 2,500 years:

> The radicalism of natural selection lies in its power to dethrone some of the deepest and most traditional comforts of Western thought, and particularly the notion that nature's benevolence, order, and good design prove the existence of an omnipotent and benevolent creator . . . To these beliefs Darwinian natural selection presents the most contrary position imaginable, [recognising] only one causal force: the struggle among individual organisms to promote their own personal reproductive success – nothing else, and nothing higher.

5

The (Evolutionary) 'Reason for Everything': Doubt

'The central question was whether the mechanism underlying microevolution can be extrapolated to explain the phenomena of macroevolution. At the risk of doing violence to the positions of some of the people at the meeting, the answer can be given as a clear, no.'

Report in the journal Science *on a conference held at Chicago's Field Museum of Natural History, 1980*

Science is by definition a progressive enterprise, hence its onward and upward march over the past four hundred years. The main impulse for its thirst for discovery is the inquisitive human mind, drawn like moths to a light to seek out and investigate the anomalies and inconsistencies of prevailing theories, in the anticipation that further investigation will reveal new facts and observations that will open the way to new and better theories. But Darwin's grand theory of evolution would remain virtually unchanged in principle for 150 years, proving impregnable to critical evaluation of its 'difficulties'. Its critics, from Richard Owen onwards, seen off with the accusation that they were seeking to reassert the theological interpretation of the wonders of the natural world; its faults set aside by its proponents, from Thomas Huxley onwards, insisting that it is better to have a speculative theory than no theory at all. Time and again this combination of resilience and elasticity would allow it to accommodate all serious challenges to its credibility. They would come in two waves. The first, in the early part of the twentieth century, arose from the discovery by the Augustinian monk Gregor Mendel of the genetic basis of inheritance that would, for a while, leave Darwin's theory without a mechanism for evolutionary change. And then, from the 1970s onwards, the difficulties posed by explaining away those troublesome twin axioms of the 'puzzle of perfection' (such as the eye) and the failure of the fossil record to demonstrate the 'continuity of life', would re-emerge, threatening 'the crust of conventional (by now, evolutionary) opinion' that had prevailed for so long.

DOUBT: PART 1

Gregor Mendel and a (Temporary) Eclipse

Darwin's imaginative leap in *The Origin* was more audacious yet than might appear, for in 1859 the practicalities of genetic inheritance, the central issue for any theory of evolution, remained quite unknown: how physical characteristics are passed on from parent to child; how genetic mutation might give rise to those 'favourable' variations; and even the details of what happened at the moment of conception. Instead Darwin supposed, as the prevailing view held, that parents passed on their characteristics to their children via 'gemules', consisting of cells budding off from the parental heart, lungs, brain and limbs which diffused into the bloodstream to be transported to the reproductive organs. Then, at the moment of conception, these 'gemules' in the paternal sperm and the maternal egg bonded together, passing on the properties of the heart, lungs, brain etc. to their offspring. Thus a cat's 'gemules' give rise to kittens, dog 'gemules' to puppies, and human 'gemules' produce babies.

In 1856, three years before the publication of *The Origin*, a Bavarian monk, Gregor Mendel, initiated a systematic study of the pattern of heredity in peas in his monastery garden that would overthrow this (to modern eyes at least) bizarre theory. Over a period of ten years Mendel laboriously pollinated by hand each of twelve thousand plants selected for their distinct character traits of *size* (tall or short), *colour* and *shape of seed*, dusting the pollen from a single, known individual plant onto the flowers of another. His findings were unequivocal: those traits (tall or short, smooth or wrinkled seeds) are transmitted as *fixed* particles (or genes, as they would subsequently be called) by parents to offspring, and on to the next generation, and so on. The parental traits do not, as the gemule theory supposed, blend together, becoming mixed and diluted. The offspring of a yellow pea and a green pea was not a mixture of the two, but either yellow or green, just as the child of a blue-eyed mother and a brown-eyed father does not have bluish-brown eyes: they will be either blue or brown.

This 'fixity' of genes explains how, for example, the same striking characteristic – such as the protuberant chin of the Spanish branch of the Austrian royal family, the Habsburgs – might be passed down unchanged through several generations. It explains, too, the astonishment so often expressed when comparing family photographs over time at the striking similarity in facial appearance, all the way down to such details as the positioning of moles or the 'gappiness' of teeth. It also explains how characteristics such as height, hair colour or a personality trait can skip a generation to reappear in the next.

Mendel duly presented his findings to the baffled members of the Natural Science Society in Brünn, and soon after was elected Abbot of his monastery. He wrote a couple of learned papers that went unread, and never performed another experiment. When he died in 1884, aged sixty-one, the budding composer Leoš Janáček was the organist at his funeral. Sixteen years later, three prominent biologists independently rediscovered his findings, which, it was immediately obvious, posed a serious threat to Darwin's theory of evolution. The concept of the 'fixity' of the gene passed on as a hard particle (like a piece of shot) through the generations clearly contradicts Darwin's supposition of natural selection acting on numerous small variations, 'each profitable to its possessor'.

Further investigation revealed that exposing flies to high doses of radiation, so as to damage their reproductive organs, generated sudden and dramatic physical changes in their offspring. Certainly these major genetic mutations were almost invariably harmful, but it was possible to imagine how a beneficial one might initiate the necessary substantial anatomical changes that could transform a fish into a reptile. Such naturally occurring, sudden and dramatic mutations seemed to offer a much more plausible mechanism for those major evolutionary transitions than supposing that they were the consequence of the 'accumulation of small inherited modifications'.

It is difficult to convey the significance of this without going into too much detail, but the crucial point is that the experimental investigation of genetic inheritance, with all its possible ramifications for the understanding of the causes of genetic disease, prospered to the detriment of Darwin's evolutionary theory. For more than twenty years Mendel's genetics cast a dark shadow over Darwin's evolutionary 'reason for everything' (or, as it now seemed to many, the 'wrong reason for everything'). But in the early 1920s, salvation would arrive from a most unlikely source – the highest level of abstract mathematics.

This is how. It would be most surprising, when out on a walk, to find that everyone was the same height and weight – say, five foot eight inches and ten stone. On the contrary, the most striking feature is people's variety: some are much taller and fatter, others smaller and thinner, and so on. But, gather together two thousand people, weigh and measure them, and an 'average' would emerge. The vast majority would fall within a narrow range of the 'mean', while the numbers of the progressively taller and shorter, fatter and thinner would fall away in a distribution known as the 'bell-shaped curve', which it so closely resembles.

The investigation of this phenomenon of the 'mean' or 'average' of physical characteristics had a special appeal for biologists with a mathematical bent, who developed statistical techniques that could define this hidden propensity of the traits of large numbers of people to tend 'towards the mean'. Both height and weight, as well as other physiological features, are determined (at least in part) by the genes, which would imply that they are not quite as 'fixed' as they might seem, but are sufficiently elastic to allow for a range of minor mutations that would allow some individuals to be tall, others small, some fat, others thin – the whole range of differences represented by that bell-shaped curve. And thus, argued Professor of Genetics Ronald Fisher (of London and later Cambridge University), it should be possible to reconcile Mendel's genetics with Darwin's theory of evolution, those minor genetic-based differences in physical characteristics being precisely the sort of small 'variations' that Darwin had suggested might confer some biological advantage on their possessors, and that, 'selected' by nature, would drive the evolutionary process forward. In essence, Fisher switched the locus of evolution from the individual to the species as a whole, where the high frequency of the same minor genetic mutations within a species of reptile (say) or bird or mammal would push it in one direction rather than another.

This account has tried so far to avoid technical terminology, in the hope that all readers might at least follow the gist of the issues involved. The purpose of reproducing here just a small part of Fisher's twenty-five-page statistical proof of his major work, *The Genetical Theory of Natural Selection*, published in 1930, is not to clarify his argument, but simply to convey its most salient point – its impenetrable obscurity.

Thus if out of a population of N individuals there are n_{11} homozygotes, and n_{1k} heterozygotes formed by combination with any other chosen allelomorph, the total of the values of x from the homozygotes may be represented by $S(x_{11})$, and that from any class of heterozygotes containing the chosen gene by $S(x_{1k})$. Then

$$\frac{2S(n_{11}) + \sum_{k=2}^{s}{}' S(n_{1k})}{2n_{11} + \sum_{k=2}^{s}{}' n_{1k}} = a_1$$

where a_1 may be spoken of as the average genotypic excess of the particular gene chosen. Σ is used for summation over allelomorphs of the same factor. If p_1 is the proportion of this kind of gene among all homologous kinds which might occupy the same locus, it is evident that

$$\sum_{k=1}^{s} (p_k a_k) = 0.$$

Fig 5-1: *Sir Ronald Fisher's statistical proof ('it is evident that...') of the Darwinian theory of evolution, in which he sought to reconcile the mechanism of natural selection acting on the accumulation of numerous small random mutations, with Mendel's mode of genetic inheritance acting via (for the most part) invariant genes.*

Fisher himself compared the explanatory power of his 'Fundamental Theorem' to Newton's laws of gravity, but his mathematical computations lay far beyond the comprehension of ordinary working biologists. 'If only we could have understood what he meant!' the evolutionary biologist George Price would observe four decades later. But luckily the biologists did not have to understand the maths, they only had to presume that it proved the reformulation of Darwin's theory, which is as follows. The genes are the major determinants of the form and attributes of living things. Every so often they 'mutate' (that is, as currently understood, every time the Double Helix divides and replicates itself, one or other of the coloured beads or chemicals that constitute a single gene is mis-spelt). That mutation may give rise to some novelty or other that provides the raw material to drive the evolutionary process forward. Nature, in turn, selects those advantageous novelties which if widespread in, say, a species of fish, might be sufficient to transform it into a reptile. Or, as a major biology textbook puts it: 'Mutation is ultimately the source of all genetic variation and therefore the foundation for evolution.'

Most such random mutations are rare, and the vast majority (over 99 per cent) of those that do have any effect are harmful in some way. 'Most mutations are more or less disadvantageous to their possessor,' comments evolutionary biologist Theodosius Dobzhansky, '[causing] deterioration, breakdown and disappearance of some organs.' The potentially *beneficial* mutation is not just rare, but very, very rare, and to have any effect must act in unison with numerous other similarly (very, very rare) beneficial mutations amongst fellow members of a species. The significance of Fisher's theorem is that, ostensibly, it proves this to be a viable mechanism for the evolutionary process. 'The thoughtful student,' counters biology professor James Mavor, 'will ask whether random mutations could have built up organisms as complicated as the higher plants and animals. At present, no satisfactory direct answer can be given. The long eons of time [over which the evolutionary process takes place] is not in itself an entirely satisfactory answer.'

Still, there is always safety in numbers, and Fisher's theorem prompted two other mathematically inclined biologists, John Haldane in Britain and Sewall Wright in the United States, to derive their own similarly impenetrable statistical 'proofs'. Mathematics is a persuasive and exact science, so for three eminent practitioners to come up with such powerful confirmatory evidence would seem to vindicate Darwin's theory of natural selection as the drving force of the evolutionary process. Except that when devising their proofs, Fisher, Wright and Haldane, as it later emerged, had started from different premises, employed different mathematical techniques, and came to different conclusions as to how the new, revised evolutionary theory actually worked.

> The unity achieved by the employment of mathematical methods . . . was accompanied by striking *disunity* in the interpretation of the way in which selection operated . . . So although Fisher and Wright were interested in showing how natural selection could operate . . . the unification was *not accompanied by one consistent explanation* about how that process actually took place.

Biologists do not, in general, understand higher mathematics, but these 'proofs' conferred a mathematical aura of invincibility on Darwin's theory that would form the basis for the relaunch in the 1930s of a modernised version, commonly known as the New Synthesis or neo-Darwinism. So now natural selection was re-established as the 'reason for everything', and more. For Julian Huxley, grandson of Darwin's great supporter Thomas Huxley and co-founder of that New Synthesis, speaking at the centennial celebration of the publication of *The Origin* in Chicago in 1959, it would

> enable us to discern the lineaments of the new religion that will arise to serve the needs of the coming era. There is no longer either need or room for the supernatural. The earth was not created, it evolved. So did all the animals and plants that inhabit it, including our human selves, mind and soul as well as brain and body.

The same year the Professor of Genetics at Edinburgh University, Conrad Waddington, in his contribution to a collection of essays, *A Century of Darwin*, struck a less triumphalist note. He drew attention to the intuitive lack of appeal in supposing, as the higher mathematics of Fisher had allegedly proven, that the glorious panoply of nature should be the consequence solely of random genetic mutation: 'Possibly it is true that genes are unstable entities which may sometimes alter spontaneously for unascertainable reasons. But this can hardly be the whole story. One would like to have a coherent, logical connection between new variation and something else in the organism's world.' He could only look forward optimistically to that 'eventual understanding' when the practicalities of how random genetic mutation might provide the raw material of the evolutionary process would be 'less patently incomprehensible than it is at present'.

DOUBT: PART 2

The Fossil Record, Perfection (and Homology) Revisited

The New Synthesis, buttressed by those obscure mathematical proofs, would prevail for the best part of forty years. But Nature is as it is, and obstinately refuses, despite much badgering, to conform with fallible human ideas of how it works, so uncomfortable facts have the disconcerting habit of sticking around despite the best efforts of those who might suppose they have been resolved. By 1980 the two most uncomfortable difficulties of Darwin's evolutionary theory had resurfaced more acutely than ever. They were, it will be recalled, the lack of evidence in the fossil record for the 'inconceivably great' number of transitional fossils required by a process of gradualist evolutionary transformation; and, second, the 'puzzle of perfection', the impossibility of demonstrating how a blind, 'trial and error', random process can give rise to 'organs of extreme perfection', such as the eye. What had happened?

(i) The fossils' verdict

The startling discovery so soon after the publication of *The Origin* of several fossilised specimens of *archaeopteryx*, with features of both reptiles and birds, had, as noted, enormously enhanced Darwin's credibility at the expense of 'the most eminent palaeontologists' of his day. Subsequently the search for similar 'transitional forms' would become a major focus of research. The most convincing yet to have been discovered is the order of *Therapsida*, whose bony skeleton has several features intermediate between reptiles and mammals. Most of the characteristics that might distinguish, say, a reptilian lizard from a mammalian rat cannot be preserved in fossilised form, so we can know nothing of the evolutionary transition from cold- to warm-bloodedness, from egg-laying to viviparity (live birth), from scaly skin to hair, and so on. But the fossilised skulls of *Therapsida* are undoubtedly 'transitional', where the bones that make up the reptile's lower jaw progressively shrink in size and move back towards the jaw joint, becoming eventually the small bones of the inner ear of mammals, the 'hammer' and 'anvil' that convert the resonance of the eardrum into the electrical impulses of the auditory nerve.

This sounds persuasive, so why, by 1980, had palaeontologists become (reluctantly) persuaded that the doubts of their nineteenth-century predecessors were well founded? The scant evidence, epitomised by the fossilised remains of *Therapsida*, of gradualist transformation between reptiles and mammals is swamped (literally) by the 'bigger picture', which reveals its antithesis – that the evolutionary process occurred in 'fits and starts'.

> 'For more than a century, biologists have portrayed the evolution of life as a gradual unfolding of new living things from old, the slow moulding

> of animals and plants into entirely different forms,' observes Steven Stanley of Johns Hopkins University. 'Today the fossil record forces us to revise this view as it turns out myriad of species have inhabited the earth for millions of years *without noticeably evolving*. On the other hand, major evolutionary transformations have been wrought during *episodes of rapid change*, when new species have quickly budded off from old ones. *In short, evolution has moved by fits and starts.*'

The main reason for palaeontologists' loss of faith in the orthodox evolutionary doctrine was the realisation that the most notable feature of the fossil record is that most of the time nothing happens. Successive strata of rock up a cliff face, representing tens of millions of years of geological time, yield: 'Zigzags, minor oscillations and the very occasional slight accumulation of change at a rate too slow to account for all the prodigious changes that have occurred in evolutionary history,' writes palaeontologist Niles Eldredge. 'When we do see the introduction of evolutionary novelty, it usually shows up with a bang, and often with no firm evidence that the organisms did not evolve elsewhere!'

This 'stasis' clearly contradicts Darwin's supposition of a continuous process of gradualist transformation, but by itself it might not seem particularly interesting. Its corollary most certainly is – for when species stay unchanged for tens of millions of years, this drastically reduces the time available for the evolutionary transformations that *do* occur, which must then happen both suddenly and rapidly. But how rapidly?

It is necessary here to recall that grand overview of the fossil record of the preceding chapter, and to focus on that crucial moment around sixty-five million years ago of the extinction of the dinosaurs, currently attributed to a change in atmospheric conditions caused by a meteor colliding with the earth. The small mammals left behind would, over the next several million years, diversify to become that vast repertoire of mammals (elephants, kangaroos, rabbits, etc., etc.), some of which (the bat) would take to the air, while others (the whale) to the sea. Darwin's imaginative (if improbable) supposition of the process of gradualist transformation in *The Origin* of that black bear swimming for hours and becoming 'more and more aquatic', ending up 'almost like a whale', could now be recast as follows: starting with a medium-sized mammal, the whale's presumed ancestor *pakicetus*, what, asked palaeontologist Steven Stanley, would it take for it to become a whale in twelve million years?

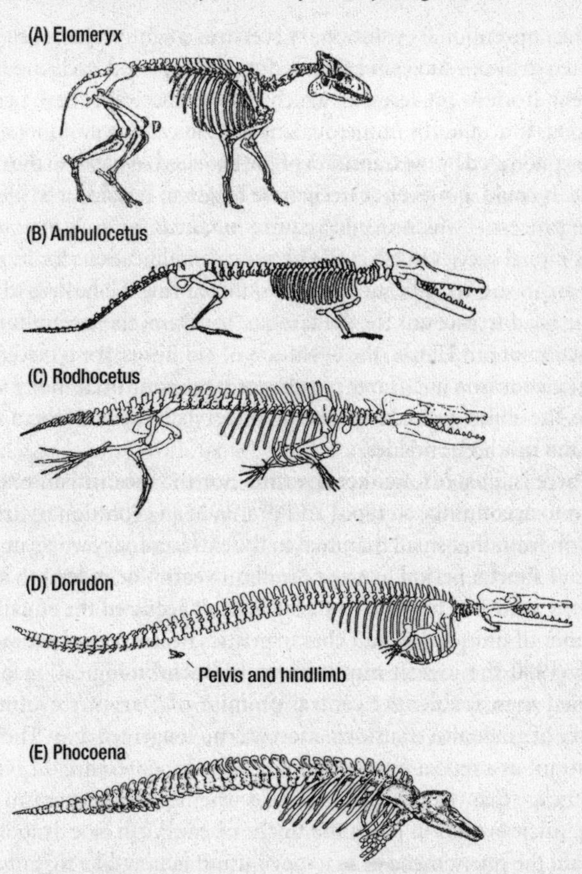

Fig 5-2: *What does it take to become a whale in twelve million years? In 1978 Professor Philip Gingerich of the University of Michigan discovered a fifty-two-million-year-old skull of the wolf-sized carnivore* pakicetus *with characteristics in common with the probable ancestor of the whale* elomeryx. *Over the following three decades palaeontologists would find the fossilised remains of the near-complete skeleton of* rodhocetus, *that swims with its legs, and the fully aquatic* dorudon, *that uses its tail for propulsion while retaining residual elements of pelvis and limbs. Together they represent 'the sweetest series of transitional forms an evolutionist could ever hope to find', observes the natural historian Stephen Jay Gould – but to fit with the Darwinian theory of gradualist evolutionary transformation they presuppose a further myriad of transitional species in the briefest of periods (relatively speaking) of geological time, of a mere twelve million years.*

The conventional evolutionary scenario might perhaps put 'end to end' ten to fifteen successive species during this period, each 'modified by descent' from its predecessor, which might conceivably have permitted *pakicetus* to acquire, by numerous small modifications, evolutionary novelties epitomised by the transition of the fox-sized *eohippus* to the modern horse. It could, however, scarcely have begun to become a whale by the same process – which would require *hundreds, or even thousands*, of trans itional species to effect the necessary modifications for its massive increase in size, modifications of the skull to bring the nostrils to the top of the head to account for the famous 'fountain', the streamlining and reduction of hind limbs, the evolution of tail flukes, the replacement of sweat glands by a thick layer of rubbery fat to control the body temperature, the ability to suckle its young underwater without them drowning, and much else besides.

There is, in short, neither the time nor the mechanism that could begin to account for so rapid and dramatic an evolutionary transformation from that small mammal to the extraordinary whale in so (relatively) short a period as twelve million years – or indeed to account for why it should have taken to the air and acquired the equally great number of unique physical characteristics necessary to become a bat.

By 1980 the logical implications of such biological 'insolubilia' seemed inescapable: the central premise of Darwin's evolutionary theory of gradualist transformation was no longer tenable. 'The central question,' as a report from a major academic conference on evolution in Chicago that year would put it, 'was whether the mechanism underlying microevolution [as in the finches' beaks] can be extrapolated to explain the phenomena of macroevolution [shrew-like mammals into whales]. At the risk of doing violence to the positions of some of the people at the meeting, the answer can be given as a clear, no.'

Some other dramatic mechanism, as yet unknown to science, must account for that extraordinary diversity of life as revealed by the fossil record. Steven Stanley comments elliptically: 'We would predict that if the ancestral species gave rise to strikingly new species this would not have happened by their own transformation, but by their budding off of *distinctive new forms*, which would, in turn, have budded off others.'

(ii) The Puzzle of Perfection

Meanwhile, the endeavours of bioengineers during the same period had thrown into sharp relief the impossibilities of Darwin's second axiom, that a succession of random variations could conceivably account for such 'puzzles of perfection' as the human eye. The many substantial contributions of bioengineers to medicine's therapeutic revolution in the post-war years include most obviously the life-sustaining technologies of the intensive care unit, such as artificial ventilation, dialysis, cardiac pacing and so on. But their attempts directly to replicate the complex functioning of whole organs such as the heart have proved far beyond their competence.

The heart is an astonishingly powerful pump, capable of propelling the entire volume of the body's five litres of blood through the 'pipeline' of arteries and veins that, stretched end to end, would circle the globe five times – 100,000 miles in all. This 'pump' may be no bigger than an orange, nestling in the palm of the hand and weighing just a quarter of a pound, but it generates enough force to propel a fountain of blood against gravity six feet into the air, and in doing so utilises as much energy as the legs of a marathon runner pounding the pavement.

The heart is not just powerful, it is also highly 'efficient', in the technical sense of performing twice as much 'work' in relation to the amount of 'fuel' utilised as any conventional man-made pump. This is due to the unique arrangement of the overlying spirals of its muscle fibres, which become progressively shorter as they taper down towards the tip (in the same way that the number of bricks in each row of a cathedral spire gradually diminishes as it tapers upwards), thus squeezing every last drop of blood out of the cavity of the ventricles at each heartbeat. And for good measure, this masterpiece of engineering efficiency should with luck run the two and a half billion cycles of a lifetime without maintenance or lubrication, or the need to replace its four sets of valves, which open and close four thousand times every hour.

Starting in the 1960s, the early pioneers of the artificial heart anticipated some difficulty in matching such specifications, but it still took twenty years to come up with the first workable device, which was implanted into a retired American dentist who went on to develop respiratory and kidney failure, pneumonia and blood poisoning before dying four months later. Over the next ten years a similar fate befell a further two hundred patients before the US Food and Drug Authority intervened and called a halt.

By now it was clear the task was hopeless – though perhaps, it was suggested, it might be possible to create an artificial heart that could act as a stopgap in patients with severe heart failure, tiding them over until a natural version in the form of a heart transplant became available. After forty years and billions of dollars, that, it seems, is as far as we are likely to get. The current model weighs twice as much, and is a fraction as efficient, as nature's version, its energy supply transmitted through a couple of tubes connected to a 'console' the size of a chest of drawers which can only be moved around on rollers and so confines the patient to hospital. This clumsy device can sustain a patient for up to two months – until a transplant of nature's much better efforts becomes available, that can keep its recipient fit and healthy for twenty years or more. In 2003, forty-two-year-old American Kelly Perkins became the first heart transplant recipient to climb the Matterhorn, having already conquered Mount Fuji (from whose summit she scattered the ashes of her donor) and Mount Kilimanjaro. She couldn't have done that with an artificial heart.

The pump-like mechanism of the heart is much the simplest of physiological systems, simpler by far than the complexities of kidney or brain, or the sense organs such as the eye. So when, as here, the purposive efforts of brilliant bioengineers employing the most sophisticated modern technology fall so far short of nature's model, it seems merely perverse to suggest that the undirected process of nature, acting on numerous small, random genetic mutations, could give rise to this or any other of those 'masterpieces of design'.

This is not to suggest there must be a Creator after all, whose higher intelligence trumped the best efforts of those bioengineers, but to draw attention to the necessity for there to be some prodigious biological phenomenon, unknown to science, that ensures the heart, lungs, sense organs and so on are constructed to the very highest specifications of automated efficiency.

(iii) The 'unsolved problem' of homology

The saga of the artificial heart resurrects William Paley's conundrum of why 'Nature's contrivances of design' should be 'greater and more to a degree that exceeds all computation' than man's much humbler efforts. Meanwhile, the findings of modern embryology would resurrect the conundrum of Cuvier's law of homology, or similarity, with its supposition of some 'formative influence' ensuring that the structures of the limbs of bats, birds, porpoises and humans were all built to the same blueprint (*see illustration, page 74*). Darwin had claimed that similarity to be powerful evidence in favour of his proposed mechanism of evolutionary transformation of one species to another. Who was right?

Darwin's interpretation of those similarly fashioned limbs as evidence of their originating from a common ancestor would, self-evidently, require that they originate from the same basic structures in the developing embryo. But it was clear from the moment nineteenth-century biologists first started scrutinising the embryos of different species down the microscope that the pattern of division of the fertilised egg, even to the untrained eye, was quite different in the major classes of vertebrates: amphibians (frogs), reptiles (lizards) and mammals (such as humans). Then, in the 1960s, the distinguished embryologist Sir Gavin de Beer, subsequently director of the British Museum of Natural History, found that the embryological origins of those homologous limbs arose from quite different 'segments' of the trunk in the newt, lizard and man.

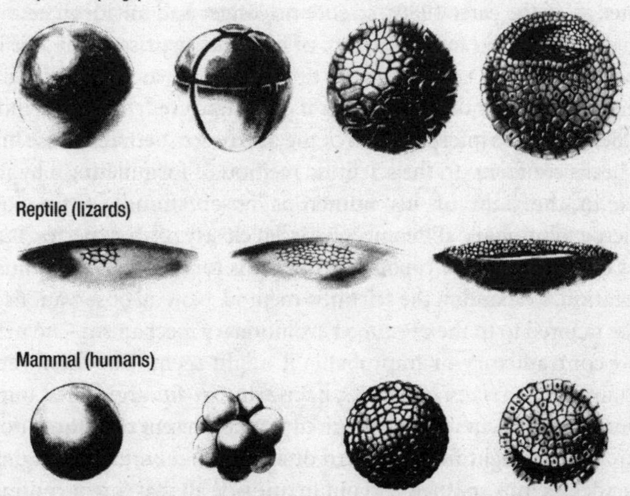

Fig 5-3: *The 'unsolved problem of homology'. At the close of the nineteenth century, microscopic studies of the early embryo revealed, as embryologist Adam Sedgwick writing in 1894 pointed out, that 'a species is distinct from the very earliest stages all through development' – illustrated here by the distinctive early forms of amphibian, reptile and mammal.*

> It does not seem to matter where in the egg or the embryo the living substance out of which homologous [similar] organs are formed comes from,' he observed in his monograph *Homology: An Unsolved Problem.* 'That [similarity of structure] cannot be pressed back to similarity of position of the cells of the embryo out of which these structures are ultimately formed.'

And if 'it does not matter', then that 'common architectural plan' of the forelimbs of reptiles and mammals, so long held to be powerful evidence for Darwin's theory, can no longer be interpreted in favour of descent from a common ancestor. 'The concept of homology is absolutely fundamental to what we are talking about when we speak of evolution,' de Beer's contemporary the biologist Sir Alister Hardy observed, 'yet in truth we cannot explain it at all in terms of present-day theory.'

Thus, as of the early 1980s, science no longer had an adequate materialist explanation for the history of life. The surprise perhaps is how readily refutable Darwin's theory turns out to be, as if it must contain some hidden flaw that invalidates its scientific credentials – as indeed it does. Darwin's interpretation of the differences between those finches' beaks conforms to the scientific method of formulating a hypothesis, in the light of his numerous observations, of the subtle 'microevolutionary' differences between closely related species. But he was obliged, when extrapolating from this form of 'micro' to 'macro' evolution, to abandon the scientific method. Now all observations had to be tailored to fit the presumed evolutionary mechanism – no matter how contradictory or improbable it might seem. The fossil record certainly appears *discontinuous*, happening in 'fits and starts', but the priority of Darwin's theory required him to present only those observations that might fit the pattern of a gradual *continuous* merging of one species into another, and put to one side all that might contradict it. Similarly, the eye with its constituent parts certainly seems as if it is designed for the purpose of seeing, but again Darwin's prior commitment to his own theory required him to argue, *per impossibile*, that it was brought into existence by the accumulation of numerous small, random variations.

Nonetheless, Darwin's theory, thanks to its resilience and tenacity, has survived virtually unscathed. Then, as now, sceptics are seen off with the charge that they are closet creationists, while its many faults are accommodated on the grounds of there being no better alternative. That tenacious preference for Darwin's doctrine remains almost universal, the supposition that he solved the mystery of the origins of life taught uncritically in schools and universities. But behind the façade of scientific unanimity, the crust of conventional opinion has been breaking up, with a dissident minority of evolutionary biologists seeking to reconcile, in a series of articles in scientific journals ('Is a New Evolutionary Synthesis Necessary?', 'Is a New and General Theory of Evolution Emerging?'), the claims of this dominant scientific theory of our age with the evidence which so clearly contradicted it. Indeed, when one camp accuses another of making 'unsubstantiated assertions resting on the surface of a quaking marsh of unsupported claims', only to elicit the counterclaim of 'ideas so confused as to be hardly worth bothering with', one might reasonably suppose that this was a theory on the point of collapse.

Meanwhile, the pioneers of the New Genetics were refining those techniques that would culminate in spelling out the genomes of fly, worm, mouse and man, whose definitive, if unintentional, implications for the validity of Darwin's evolutionary theory have already been hinted at. We turn now to that concluding episode of this great saga, whose astonishing and totally unexpected insights into the mechanisms of genetic inheritance would put to the test the 'scientific knowability' of the means by which the Double Helix imposes the order of form on the near-infinite diversity of life.

6

The Limits of Science 2: The Impenetrable Helix

'How is it we have so much information, but know so little?'

Noam Chomsky

The elegant spiral of the Double Helix, like Newton's law of gravity, combines great simplicity with phenomenal power. But the practicalities of what it does, how it imposes the order of 'form' and all the complexities of life on the fertilised egg, are of a qualitatively different order – and for the obvious reason that 'life' is immeasurably more complex than 'matter'. Compare, for example, a pebble at the bottom of a pond with a humble fly of similar size on its surface. The pebble is composed of several billion atoms of calcium and phosphorus, all neatly arranged in a highly regularised pattern. And the fly? The contrast is so startling that it scarcely needs spelling out: a similar number of atoms are here arranged in an enormously greater number of different ways – as its *face*, with antennae, eyes and mouth parts; its *dot-sized brain*, which orchestrates its flying skills; its *wings*, beautifully jointed *limbs*, and so on. To the seventeenth-century naturalist, this symphony of parts, glimpsed for the first time through the newly discovered microscope, was a revelation:

> It seems as though God has bejewelled them in compensation for their lack of size. They have crowns, blooms, and other attire upon their heads against which anything invented by the richest of men must pale; and those who have used only their eyes have never seen anything so beautiful, so fitting or so magnificent in the houses of the greatest

princes as that which can be seen with magnifying glasses on the head of a simple fly . . . so much beauty concentrated in so small a space . . .

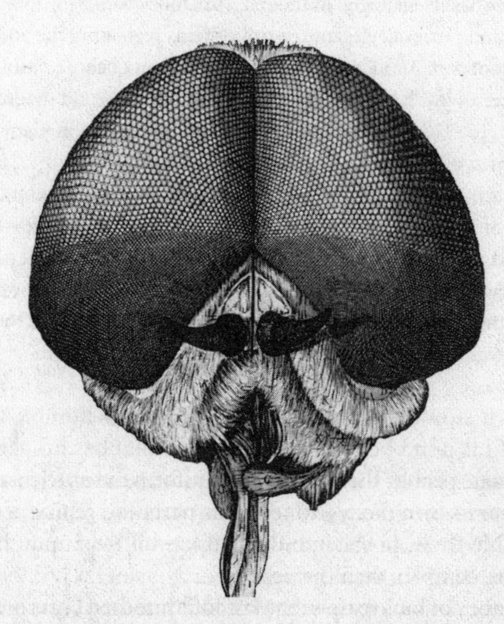

Fig 6-1: *Robert Hooke's famous drawing of the head of a fly, with its prominent 'compound' eyes and appendages, from his major work* Micrographia *(1664).*

But this contrast between the outer physical appearance of fly and pebble is merely the beginnings of the differences between them. For the fly, or any form of life, is fashioned from successively more fundamental 'organisational' levels, or layers, of increasing complexity. First, the organs such as the heart, brain and gut. Those organs in turn are fashioned from the unique arrangement of nerves, muscles and glands that make up the tissues. The tissues themselves are fashioned from the most fundamental unit of life, the cell; at which point if we were to step inside we would find ourselves 'in a world of supreme technology'.

'We would see endless corridors branching in every direction [along which] a huge range of products [proteins, enzymes] and raw materials [chemicals] shuffle in a highly ordered fashion to and from all the various assembly plants,' writes biologist Michael Denton. 'We would wonder at the level

> of control implicit in the movement of so many objects, all in perfect unison. We would see that nearly every feature of our own advanced machines has its analogue in the cell:...memory banks for information storage and retrieval, elegant control systems regulating the automated assembly of parts, proof reading devices utilised for quality control, assembly processes involving the principle of prefabrication and modular construction ... This automated factory carries out almost as many unique functions as all the manufacturing activities of man on Earth ... but with one capacity not equalled in any of our most advanced machines – it is capable of replicating its entire structure within a matter of a few hours.
>
> '[And yet] the cell, this remarkable piece of machinery that possesses the capacity to construct *every* living thing that has ever existed is several thousand million million million times smaller than the smallest piece of functional machinery ever constructed by man.'

Nor does it stop there, for the properties of the humble fly on the surface of his pond belong in a dimension that has no parallel with the inanimate pebble: the capacity to transform the nutrients on which it feeds into its own tissues, to repair its parts and reproduce its kind. The humble fly is, in short, billions upon billions upon billions of times more complex than the pebble.

The history of biology over the last four hundred years is essentially the progressive and purposive unravelling, in successive waves of discovery, of the ever more fundamental layers of the complexity of life – from the descriptive anatomy of the whole organism in the seventeenth century, through the physiology, histology, embryology and biochemistry of the eighteenth and nineteenth centuries, culminating in the penetration over the last sixty years of the mysteries of the cell and the instructions determining the form of all living things strung out along the Double Helix. There is nowhere further for biology to go. This is the end of the line, and just as the fly is so vastly more complex than the pebble, so the directive force of the genetic instructions of the Double Helix must, when compared to Newton's physical laws of matter, be billions upon billions upon billions of times more potent. The deciphering of those genetic instructions would be biology's last great task: how do they impose form on living things? How do they replicate themselves to be transmitted from one generation to the next? How might they change so that one form of life is transformed into another?

There are no profounder questions. Their elucidation over the past 150 years has come in three major instalments. It began, as already touched on, with Mendel's famous pea experiments, which revealed how the characteristic traits of living things are passed down from one generation to the next in the form of 'genes', for the most part fixed and unchanging, ensuring that the offspring of cats will always be kittens, except for the occasional – usually harmful – 'mutation'.

The second instalment began in 1953, when Francis Crick and James Watson discovered that those genes were strung out along the two intertwining strands of the Double Helix, each strand a sequence of just four chemicals that, so far, we have portrayed as 'coloured discs', but which are commonly designated by the first letters of their names: C(ytosine), G(uanine), A(denine) and T(hymine). Christopher Wills, Professor of Biology at the University of California, draws an appropriate analogy:

> I have just with some difficulty hefted *Webster's Third New International Dictionary* onto my lap. This is the one that you see in libraries, sitting proudly on its own little lectern. I find there are about sixty letters to a line, 150 lines to a column and three columns to a page. This works out at 27,000 different letters to a page. The dictionary has roughly 2,600 pages adding up to seventy million letters in all. Since there are about three billion [letters] in the human genome, it would take forty-three volumes the size of this enormous dictionary, filling a shelf twelve feet long, to list the information that they carry.

In reality, of course, those C G A T chemical molecules are minuscule compared to the letters representing them, so when strung together along the Double Helix they produce a continuous strand of approximately three inches (or seventy-five millimetres) long. But those seventy-five millimetres must be packed into the nucleus of the cell, which is about 1/500th of a millimetre across. The Double Helix, in achieving this extraordinary feat, 'packs over a hundred trillion times as much information by volume as the most sophisticated computerised information system ever devised'. This is how it works.

We start by straightening the helix to form a ladder whose outer banisters (as it were) are formed from two continuous strands of sugar molecules (known as deoxyribose). From each is suspended the sequence of C G A T chemicals (the crucial part of a compound known as nucleic acids), that grip together in the middle to form the ladder steps – three

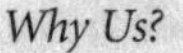

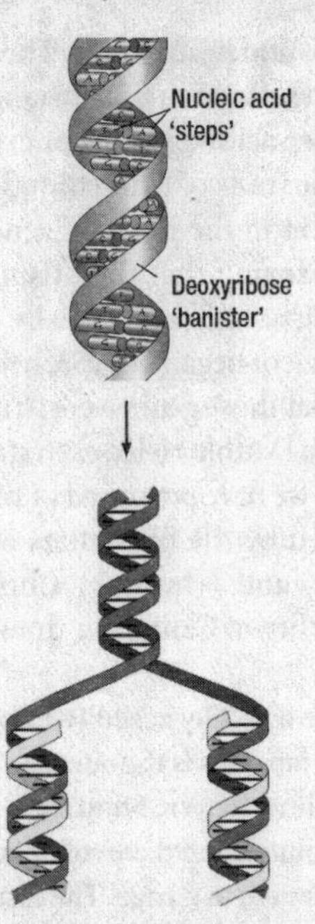

Fig 6-2: *The familiar schematic representation of the process by which the Double Helix of DNA replicates the genetic information every time the cell divides by 'the formation onto itself of a new companion chain so that eventually we shall have two pairs of chains where we only had one before'.*

billion in all. The initials of the Deoxyribose sugar 'banisters' and the Nucleic Acid chemicals of the 'steps' give the technical term DNA.

This memorably simple and elegant structure is perfectly suited to the necessity of replicating the 'trillion times compressed' information every time the cell divides to form two daughter cells. First the Double Helix splits down the middle, and the stubs of the half-steps on either side each now form a complementary sequence of letters. These stubs then form the template on which the cell builds a further complementary strand – so one Double Helix becomes two, or, as Crick and Watson described it in the 'most important scientific paper in the history of biology':

> We imagine that, prior to duplication, the bonds [connecting the two half-steps] are broken and the two chains unwind and separate [the ladder, as

> it were, splits down the middle], each chain then acts as a template for the formation onto itself of a new companion chain so that eventually we shall have two pairs of chains where we only had one before. However, the sequence of the pairs will have been duplicated exactly.

Crick and Watson's economical prose does not begin to do justice to the staggering logistics of this process: with up to four million new cells being created in the body every second, a thousand enzymes swarm down the Double Helix like a team of readers systematically ploughing through the volumes of Webster's Dictionary, copying each of the three billion C G A T chemicals in precisely the right sequence – and then stringing them all together again.

We turn now to the crucial question of *how* the Double Helix embodies 'information', or specifically, how those C G A T molecules are arranged into genes, and how those genes 'instruct' for the complexities of life. Here the serial arrangement of those molecules obviously represents some form of 'code'. This is best understood by referring back to that description of the workings of the 'automated factory' of the cell. We will focus on just one of the 'products' as it rolls

Fig 6-3: *The bafflingly complex haemoglobin molecule consists of four intertwined chains, each made up of 140 'parts'. These molecules, packed in their hundreds of thousands within a single red blood cell, convey oxygen from inspired air in the lungs to the tissues of the body to drive the great cycle of life.*

off the assembly line – the haemoglobin molecule, tens of thousands of which in each red blood cell capture the atoms of oxygen in the inspired air of the lungs and transport them to the nerves and muscles, to fuel the chemical reactions that drive the machinery of their cells.

We rightly marvel at the bewilderingly intricate structure of this haemoglobin molecule which performs so simple a task – but it is no more intricate than the tens of thousands of other 'products', proteins and enzymes rolling off the cells' assembly lines from which all living things are made. We find, scrutinising that haemoglobin molecule more closely, that it is fashioned from hundreds of small parts, which on closer scrutiny still come in just twenty different forms – comparable to the rectangles, squares, circles and triangles in a box of children's bricks. These different 'parts' (known as amino acids) come from the food we eat, digested in the gut and transported to the cells to be reconstituted into ourselves and products such as haemoglobin. Now, reverting to the genetic 'code', we find that those C G A T chemicals strung out along the Double Helix, when arranged in threes (or triplets) – say, C G G – each 'speak for' one or other of those twenty different-shaped parts.

This may be the point at which some might fear losing the thread – but with a simple diagram (*opposite*) all should become clear. How does the red blood cell make those haemoglobin molecules? First, the Double Helix splits at the point where several hundred of those C G A T letters, arranged in triplets, form the haemoglobin 'gene', each triplet 'speaking for' one or other of those parts. The gene makes a duplicate of itself, known as its 'messenger'. The 'messenger' then passes out of the nucleus into the main part of the cell to find one of the many mini-factories for making the 'products' of the cell. The 'messenger' feeds itself, like a tickertape, in at one end of the mini-factory, reads the first triplet, and finds and adds the relevant part, one of those different-shaped building blocks, to the assembly line. Then it reads the next triplet (ditto), then the next (ditto) and the next (ditto), until the last triplet is reached and the final part is in place. And then what?

Those different-shaped parts (squares, circles, triangles, etc.), now strung out in a long, thin sequence, slip off the assembly line and, hey presto, all spontaneously 'spring together' to form that bafflingly complex haemoglobin molecule. By analogy: suppose you wish to buy a new home-assembly chest of drawers for the bedroom. You purchase its component parts from a furniture store, lay them on the

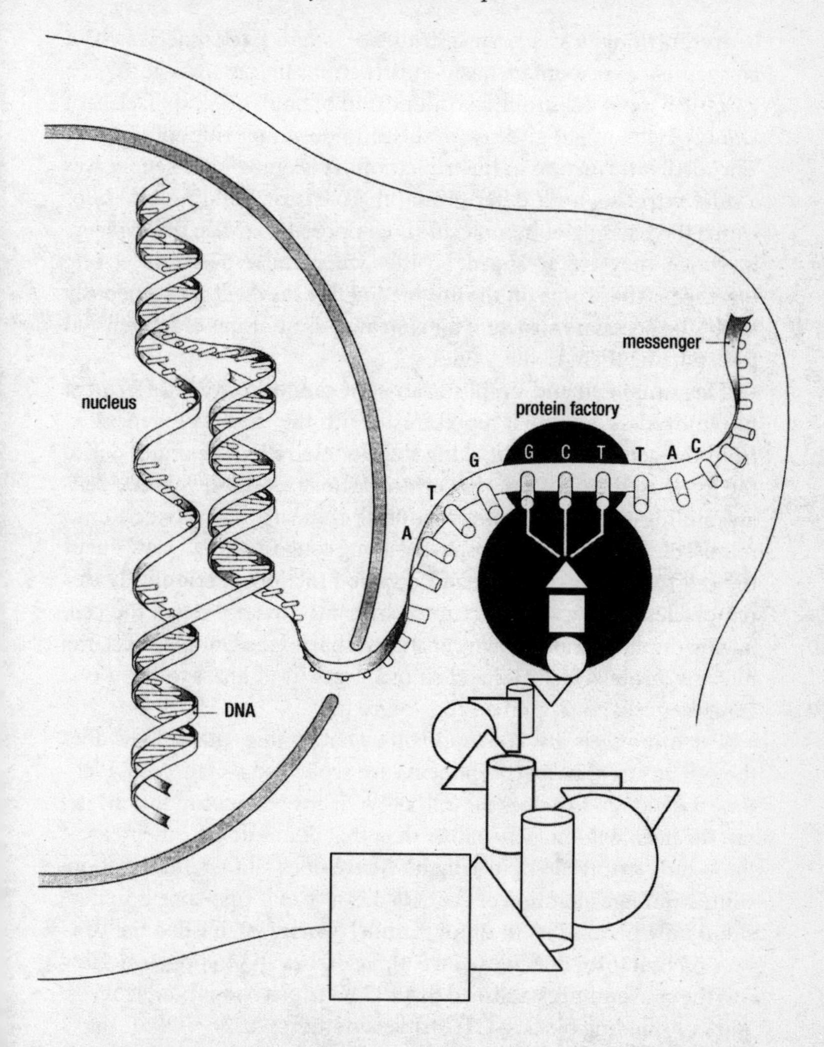

Fig 6-4: *How it works. The blood-forming cells of the bone marrow first create a copy of the haemoglobin gene which then passes out of the nucleus into the cytoplasm of the cell. This copy then feeds itself, like a tickertape, into one of the thousands of protein factories (or ribosomes), its coded information ensuring that the parts of the haemoglobin molecule (represented here by circle, square, triangle, etc.) are selected and placed in the right order on its production line.*

bedroom floor in the correct sequence, stand back, and watch in amazement as it spontaneously constructs itself. It is possible to get a glimpse here of the stunning conundrum of how the 'simple', elegant Double Helix might give rise to the infinite complexities of life. The 'encoded' information in the triplets of the haemoglobin gene serves a dual purpose, both determining the parts or building blocks of which the haemoglobin molecule is composed, and also the *sequence* in which they are arranged, so that when those parts are strung together as they come off the production line they will spontaneously adopt the necessary structure for clutching those atoms of oxygen and transporting them to the tissues.

This simplified and simplistic account cannot convey the ferment of remorseless activity it represents – with the 'messages' generated from the genes strung out along the Double Helix streaming out of the nucleus; the parts pouring off the factory assembly line and self-assembling themselves into 'products'; these are then loaded onto minuscule 'trucks' to be transported along congested 'highways' out of the cell to fulfil their many and diverse functions. Periodically this remorseless activity is interrupted by a massive upheaval; the cell, having duplicated all its own constituent parts (the Double Helix, the nucleus, protein factories and so on), tears itself apart to form two 'daughter cells', each a direct copy of itself.

The mind reels just contemplating what all this entails. How does the cell ensure that its components are replicated so faithfully every time it divides? How does the cell 'know' at any one instant to activate, say, the haemoglobin gene rather than that of any of the other tens of thousands strung along its length? Where does all that 'information' come from: the intention of the SOS message in Morse code is understood only because its inventor, Samuel Morse, established the convention that three dots mean an S, three dashes an O, and so on. How was the convention established that a GAG triplet should code for one 'part' or building block, a GTG triplet another?

The Double Helix, to do so much, possesses two sets of extraordinary, and indeed contradictory, properties. First, it is both *stable* and *dynamic*: 'stable' in that it passes on precisely the same genetic information from generation to generation over millions of years; yet also 'dynamic', for the vitality that distinguishes the biological world from the physical (a pebble from a fly) is the dynamic capacity to make a replica of itself every time the cell divides.

Next, the Double Helix is both an *architectural blueprint* and a *skilled craftsman* in one: it both contains the plan for the individual organism, and directs its development down to the last detail. These dual powers confer on the Double Helix an almost spiritual dimension. It is immortal, being passed on from generation to generation, and yet for each individual it determines who they are, their physical and mental attributes, and what they will become.

> 'On the one hand [there is] the simplicity of the principle, on the other the endless complexity of the process,' writes biologist Maitland Edey. 'It is that magisterial power that commands our attention. Its four molecules are absurdly simple. They are not alive, but they can do things no [molecules] ever dreamed of . . . They wave their instructional wands . . . to make everything that can be called alive, to monitor the development of every fern and feather on earth, to direct their growth, to enable them to function, to replace worn-out parts, to turn things on, to turn them off . . . Lo, there emerges a bacterium, a flower, a fish, a Frenchman.'

The Double Helix is an extraordinary phenomenon. But, tantalisingly, the elegant simplicity of its intertwined strands held out the promise that those extraordinary properties and the prodigious amount of information they contained might be decipherable – that the fundamental mysteries of 'life' might be *knowable*. And that, of course, was precisely the promise of the third and final instalment of the unravelling of the Double Helix, those technical innovations of the 1970s, touched on in the opening chapter, which held out the reasonable expectation that spelling out every gene would reveal *the true cause* of living things, and how chance changes or mutations in those genes could bring about their evolutionary transformation one into the other.

It is difficult to convey the excitement amongst biologists as they realised how those three technical innovations of the New Genetics might permit them to read and interpret the genetic instructions coded within the Double Helix. The first of those innovations, it may be recalled, permitted researchers to cut up the sequence of the three billion chemical molecules, C G A and T, that made up the Double

Helix, into manageable fragments. Next they learned how to produce numerous 'photocopies' of those fragments, the better to find out what they did. And finally they developed the methods that would 'spell out' the sequence of letters or chemicals of the individual genes, and thus work out which products they coded for and what they did.

This was real science, of the sort that fills the pages of learned journals and earns Nobel Prizes all round; and with the forty-three Webster-sized volumes of information waiting to be explored, there was more than enough for everyone to be getting on with. Surprisingly, the daunting prospect of working it all out soon appeared much easier than expected, and for the most unlikely reason – the realisation that 95 per cent of that Double Helix was 'junk', tens of thousands of repetitions of just one letter of the genetic code (an A say, or a G) that, not coming in the form of threes or triplets, could not 'speak for' one or other of those parts or building blocks. To clarify. Nature is invariably parsimonious – a fly has all it takes to be a fly, nothing more, nothing less, a worm to be a worm, and so on. It would therefore be reasonable to expect to find the genetic instructions strung out along the Helix to be expressed in some similarly cogent and compressed way. This is the supposition behind Christopher Wills' analogy of there being forty-three volumes of 'information' packed within the nucleus. But that is not how it turned out. Instead, the disproportionate amount of 'junk' in the Double Helix means that the equivalent of forty-two of the forty-three Webster volumes contain *no genetic information at all*, while the genes themselves come in fragmented pieces, separated out again by further long sequences of 'junk' that first have to be 'edited out' to form a coherent message. Put another way, the arrangement of those genetic instructions along the Double Helix is almost the reverse of what one might reasonably expect. There is, after all, little difficulty in providing the directions to your house (turn left at the post office, past the petrol station, etc., etc.) – as long as you do not chop the words up into little pieces and intersperse them with page after page of meaningless gibberish.

This was the first intimation of the perplexities to come. There was every reason to suppose that the profound complexities of life would require an equivalent prodigious amount of 'information', but the surprise was how *little* information was required, with just one 'volume's worth' of Webster's International Dictionary containing sufficient instructions to transform a single fertilised egg into a human being. It was all very odd.

There was, however, no difficulty in putting such oddities to one side, for right from the beginning the techniques of the New Genetics proved spectacularly successful not just in tracking down and identifying thousands of genes, but in identifying the changes or 'mutations' in the genetic instructions that result in all types of genetic illnesses – of which the simplest to understand is the mutation in the haemoglobin gene that causes the blood disorder sickle-cell anaemia. This, in turn, provided the first opportunity critically to examine that central concept of the evolutionary theory, how such random 'mutations' might also give rise to beneficial variations, which cumulatively might create those organs of extreme perfection such as the eye, and cause the transformation of one species into another.

The red blood cells in those who suffer from sickle-cell anaemia, as the name suggests, are 'sickle' shaped, causing them to clump together, blocking the bloodflow through the capillaries and depriving the tissues of oxygen – with painful and potentially lethal consequences. Back in 1956 an English chemist, Vernon Ingram, working at the famous

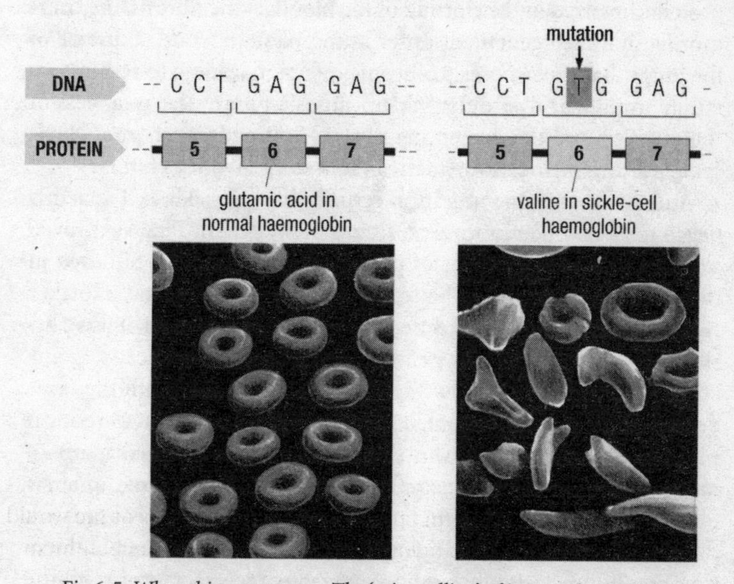

Fig 6-5: *When things go wrong. The 'mis-spelling' of just one letter of the haemoglobin gene ('T' instead of 'G') causes the protein factory to insert the 'wrong part' ('valine' rather than 'glutamic acid'), with catastrophic effects for the shape and functioning of the red blood cells.*

Cavendish laboratory in Cambridge (where Crick and Watson had discovered the Double Helix just three years earlier), found the instability of the red blood cells to be due to a faulty arrangement of just one of the parts of which the haemoglobin molecules are made. Two decades later, when it became possible, thanks to the New Genetics, to spell out the full sequence of the haemoglobin gene, it emerged that that faulty arrangement was caused by a single mutation which should normally read GAG, but is mis-spelt in those with sickle-cell anaemia to read GTG – so the protein factory inserts the 'wrong' part. By analogy with that self-assembling chest of drawers, just a single defective part would cause it to collapse.

This first demonstration of the devastating consequences of just a single 'mis-spelling' or mutation could scarcely be more persuasive, and before long the genetic mutations responsible for many important disorders had been identified: the bleeding disorder haemophilia, famously transmitted by Queen Victoria to the royal households of Europe, caused by a mutation in the gene that codes for one of the proteins involved in the clotting of the blood; cystic fibrosis, the commonest inherited genetic disorder in the Western world, a disease of the lungs that predisposes to chronic infection leading to respiratory failure; muscular dystrophy, which causes a progressive weakness of the muscles; and the dementing illness Huntingdon's chorea, whose many victims include the American folk singer Woody Guthrie.

And yet, that single mutation responsible for sickle-cell anaemia began to appear ever more exceptional. Many genetic illnesses proved to have not one but dozens of possible 'mutations': two hundred in the case of cystic fibrosis that could cause the disease, and a further two hundred that made no difference. Further, it emerged, it was possible to have a devastating genetic mutation without it causing any abnormality at all – as illustrated by two sisters, both with the same defective gene for the inherited form of blindness known as retinitis pigmentosa. The younger was indeed blind, but the visual acuity of her older sister, whose gene contained exactly the same mutation, was excellent, and did not prevent her from working as a night-time truck driver. These complexities, inexplicable in the conventional understanding of how genes behave, became ever more perverse, with the same genetic disease caused by different mutations in several genes, while, contrariwise, several different diseases can stem from mutations in a single gene.

The puzzle proliferated further when biologists, seeking to define more precisely the function of those newly discovered genes, devised some ingenious experiments that involved 'knocking out' the gene in the embryo of an experimental animal (such as a mouse), and observing the effects on its subsequent development.

> 'In some cases a "knockout" may have no effect whatsoever, despite the fact that the gene codes for a protein that is thought to be essential,' writes Michel Morange, Professor of Biology at Paris's École Normale Supérieure, describing the 'discouraging' result of experiments involving more than a thousand separate genes. 'Sometimes the knockout can even improve the performance of the mutated animal(!)...or have effects that are completely different from those that were expected. Then the inactivation of the gene does indeed produce the predicted defect, but to a much lesser extent than expected . . . Finally, there are cases where the inactivation of a gene does indeed lead to the predicted effect, but these examples are relatively rare.'

These 'discouraging' results required a drastic shift from the naïve (in retrospect, at least) notion that the individual genes 'for this and that' might explain the complexity of living things. Rather, it began to appear that the genes did not have single, discrete functions, but are 'multi-tasked', the same gene being involved in the development of many different parts – eyes, nose, brain, pituitary gland, gut and pancreas – and not alone, but in unison with thousands of other genes all working together: six thousand of the fly's total complement of thirteen thousand genes are involved in the formation of its heart, while the sexual organs of a worm less than a millimetre in length are guided by the interaction of two thousand genes. Further, it is now clear that 'context is all'; so, depending on what is required, the gene can have quite contradictory properties, in one context promoting the growth of the cell, but in another its self-destruction. Similarly, the same gene can be both essential for, and yet irrelevant to, its purpose: the much-studied gene involved in the formation of eyes is also found in blind worms, while 'knocking it out' in another species seems to have no deleterious effect, as the eyes form nonetheless.

The elegant simplicity of the Double Helix began to seem ever more deceptive, as it now appeared that the genes must somehow be acting in unison in staggeringly complex ways when instructing those factories in the cell to turn out their 'products' that form the complex tissues of the brain, heart, lungs and so forth.

> 'The heart of the problem,' observed Philip Gell, Professor of Genetics at the University of Birmingham, 'lies in the fact that we are dealing not with a chain of causation but with a network, that is a system like a spider's web in which a perturbation at any point of the web changes the tension of every fibre right back to its anchorage in the blackberry bush.'

And that 'spider's web' seemed to pose an ever more daunting challenge for the supposed mechanism of evolutionary transformation. For when it takes six thousand genes to build a heart, what chance was there that a 'random mutation' in any one of them might generate a beneficial variation in favour of the heart's further perfection? Perhaps there were some 'mastermind' switching genes, turning the others 'on and off' according to some preconceived plan. Find those master genes, and the perplexing and contradictory findings would become clear. And sure enough, in the late 1980s, soon after Philip Gell's cautionary analogy with the spider's web, the Swiss biologist Walter Gehring discovered two clusters of those master genes. These Hox genes, as they are known, determine the three-dimensional organisation of the front and back half of the fly respectively, the first cluster being involved in the development of the antennae, legs, wings and so on, while the second does the same for the 'tail end'. Perhaps those master genes might finally reveal the secret of those evolutionary transformations – where (and why not?) some random mutation might change the form of one species into another; not gradually, as Darwin had surmised, but suddenly and dramatically causing a fish to move to dry land and become an amphibian, or a land-based animal to sprout wings and become a bird, or a shrew-like mammal to take to the sea and transform itself into a whale – just as the pattern of sudden emergence of the fossil record suggests.

But when Gehring and his colleagues pursued this extraordinarily important discovery further, they found something yet more astonishing still (indeed, probably the most astonishing discovery in the history of biology): that precisely the *same* 'master' genes mastermind the three-dimensional structures of *all* living things: frogs, mice, even humans. The same master genes that cause a fly to have the form of a fly, cause a mouse to have the form of a mouse. 'The central significance' of such findings, observed the late influential biologist and writer Stephen Jay Gould, 'lies not in the discovery of something previously unknown – *but in their explicitly unexpected character*'; to which we now turn.

From the mid-1990s onwards, the powerful techniques of the New Genetics would culminate in the genome projects, in anticipation that spelling out the full complement of genes would reveal the Holy Grail of biology, the secrets of genetic inheritance that distinguish one form of life so readily from another. But the findings of those genome projects would, as touched on in the opening chapter, prove yet more perplexing still than all that had gone before – the surprising *paucity* of genes for the task of fashioning the diversity and form of living things, and their near-*equivalence* across so vast a range of complexity, from bacteria to humans. Nonetheless, the genomes, taken together, encompassed the entire evolutionary trajectory, from the simplest, earliest forms of life (bacteria) to the most complex (humans), and so, comparing one with another, might account (at least in part) for those apparent perplexities.

It is possible to get a feel for that trajectory at an elementary (but instructive) level by tracing the progressive increase in the size and complexity of the full complement of genes, or genomes, over time. We start with the earliest, smallest and simplest form of free-living organism, the bacterium, little more than a drop of interacting chemical reactions surrounded by a cell wall whose fossilised remains first appear in strata of rock laid down 3,500 million years ago. Bacteria might seem simple, but being billions of times more complex than inanimate matter (such as a pebble), are in reality not simple at all. Their genomes range in size from the 470 genes of the (pneumonia-causing) *mycoplasma*, presumed to be the irreducible minimum necessary for an organism to function and reproduce its kind, all the way to the eight thousand genes of the soil bacterium *Mesorhizopium*, that promotes the growth of plants.

It would take another 1,500 million years for the next step in that evolutionary trajectory, the emergence of strikingly more sophisticated single-cell organisms such as the yeast *Saccharomyces*, whose extraordinary properties we glimpsed in 'that world of supreme technology' outlined earlier. The genomes of these more complex forms of free-living organism are on average twice as large – coming in with a genome count of just under five thousand.

And the next step after *that*, after a further interval of 1,500 million years, would be the dramatic Cambrian explosion five hundred million years ago of 'multi-cellular' forms of life, with the emergence of their several distinct body plans together with the necessary prerequisites for their survival: the sense organs of sight and hearing to perceive the world around them, digestive and circulatory systems, sex organs for reproduction, and so on. This leap in organismal complexity would, not surprisingly, have required a massive increase in the complexity of the genome, where even the blind, millimetre-long roundworm *C. Elegans*, with just 959 cells in total, comes in at 19,100 genes. It is thus more than surprising to discover that the prodigiously more talented fly, with its eyes, wings, legs and capacity for memory and courtship, should turn out to have a genome of just 13,600 genes, while we humans, our primate cousins and mice get by with around twenty-five thousand genes, which would seem scarcely sufficient to begin to account for, for example, the human brain with its tens of billions of nerve cells.

This brief history of the evolution of life as revealed in the landscape of these genomes confirms an underlying trend of a progressive increase in gene numbers from bacteria through yeast, flies, worms, mice and ourselves – but that history offers not the slightest hint of the driving force behind the ever-increasing complexity of the information encoded by those genomes, nor how they might account for those characteristic features that so readily distinguish one form of life from another. Their genomes and those of all the organisms with which we share this planet do no more than code for those same 'products' or parts of the cells that are common to all, and from which all living things are formed. Thus we share, unchanged over thousands of millions of years, many of the genes found in bacteria, for they, like us, utilise the same enzymes that drive the chemical reactions that generate the energy that keeps them (and us) alive. We share many more of our genes with the fly and the mouse, because their 'housekeeping' duties are the same. They too require haemoglobin to transport oxygen from the lungs to the tissues, insulin to ensure a constant supply of sugar in the blood, and reproductive hormones that will ensure the propagation of their kind. So the seeming perplexity of the *economy* and *equivalence* of genes across so vast a range of life is, at this level at least, explained. How is it then that living things are so unambiguously distinctive one from the other?

Hence the significance of those 'master' or regulatory genes just touched on, which though constituting a fraction of the total (around 2 per cent), nonetheless possess the capacity to conjure from that universal 'gene kit' the antennae, legs and wings of a fly. But the relief at their discovery, as noted, was short-lived, when it emerged that the same master genes that cause the fly to be a fly, cause a mouse to be a mouse. The implications of this conundrum are well illustrated by the eyes respectively of flies and mice, which are 'constructed' on a very different plan. The mouse eye (like our own) is the familiar camera type that so prompted the admiration of William Paley. By contrast, flies have 'compound eyes' composed of sheets of lenses at different angles. But the same 'master gene' known as Pax 6 brings both forms of eye (and indeed all eyes) into existence.

This *directive* power of the same Pax 6 gene in the formation of these very different types of eye is further illustrated by two of the most astonishing experiments in the history of biology. In the first, scientists

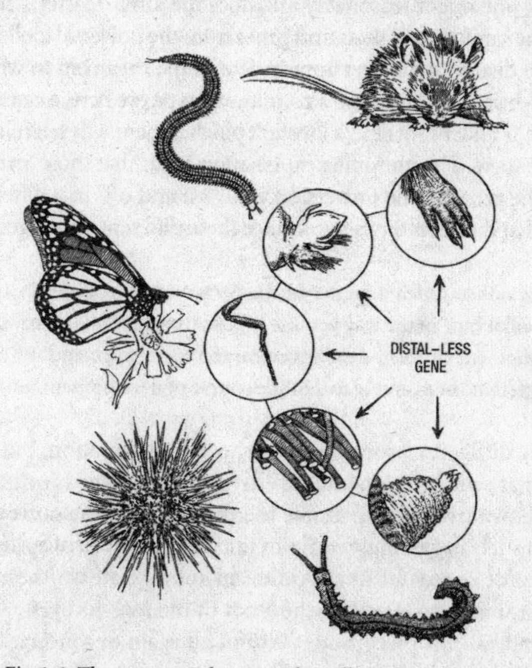

Fig 6-6: *The same gene, known technically as distal-less, orchestrates the formation of the diverse limbs of mouse, worm, butterfly and sea urchin.*

ingeniously activated the Pax 6 gene in many tissues of the embryonic fly – causing it to form eyes on its wings, legs and other parts. Next, the experiment was repeated, but this time by introducing the mouse version of the gene, which in the mouse gives rise to the camera-type eye, but which when inserted into the fly embryo gives rise to those extra fly-like compound eyes. And so too the limbs, where the same gene that orchestrates the formation of a fly's leg also instructs for the limbs of crustaceans, spiders, centipedes and chickens. And the heart too: 'Along the topside of the fly there is a heart that contracts to pump fluid around the inside of the body. It is not much of a heart by human standards but it does the job,' writes Professor Sean Carroll of the University of Wisconsin. 'Geneticists discovered the gene required for making the fly heart, and named it Tin Man (after the character in *The Wizard of Oz* who lacked a heart).' But Tin Man, it emerged, also played an important role in the formation of the infinitely more complex and differently structured four-chambered human heart.

How then, one might reasonably ask, does the same 'master switch' orchestrate the same several thousand genes from the universal toolkit to produce such diverse structures appropriate to the organism to which they belong – bringing into being a compound-type eye here, a camera-type eye there, a spider's leg here, a lobster's pincher there, a fly heart here, a human one there. The prevailing explanation holds that those 'master genes' turn the genes of the universal toolkit 'on and off' in a different sequence and at different times to produce these different structures.

> 'Diversity is not so much a matter of the complement of genes in an animal's toolkit but "in the way you use it",' observes Professor Carroll. 'The development of form depends upon the turning on and off of genes at different times and places in the course of development.'

It is certainly difficult to conceive of any other explanation, but the logistics of that switching 'on and off' defy all imagining as one tries to conceive how it would be possible to generate from the same universal toolkit such staggeringly complex and distinct types of eye, each so beautifully 'designed' for its purpose and appropriate for the creature to which it belongs. Further, the parts of the fly – its eyes, wings and limbs – are all 'of a piece', and it is difficult again to conceive how the relevant master gene for each could have chanced upon the correct sequence of switches to generate the appropriate part. It is as if the 'idea' of the fly (or any other organism) must somehow permeate the genome that gives rise to it, for it is only through the master genes of the embryonic fly *knowing it is a fly* that they will activate that sequence of switches that will give rise to those appropriate structures.

There is clearly a problem here. The genomes strung out along the Double Helix *must*, we know, give rise to the infinite diversity of form and attribute of living things, all the way down to the last detail of every species of tree and plant, fish, bird and animal, and humans – otherwise those details could not be passed down with such fidelity from one generation to the next. Why then does this seductive simplicity of the Double Helix contain not the slightest trace of those details? Clearly the Double Helix is not simple because it *is* simple. It is simple because it *has to be* simple if it is to pass on 'the code of life' by duplicating itself every time the cell divides. And that *obligation* to be simple requires the Double Helix to condense within itself all those complexities that would distinguish one form of life from another.

The implications are twofold. First, the necessity for the Double Helix to be simple means that it cannot act alone. Rather, the vastly more complex cell within which it is located must somehow 'know' its own needs, and then determine which of those 26,000 genes at any moment are to be activated to send the appropriate tickertape message to the protein factory to produce one or other of those thousands of parts. And the cell in issuing those instructions will in turn be influenced by the needs of the tissues and organs within which *it* is located, and so on all the way up the hierarchy to the organism in its entirety. Next, that simplicity also requires that the principle of self-organisation (epitomised by the haemoglobin molecule spontaneously assembling itself from the parts of which it is made) must also apply at successively higher levels of 'organisation' from the automated factory of the cell upwards. Further, from an evolutionary perspective, while the near-equivalence of the human genome to that of our primate cousins would certainly confirm our 'common ancestral heritage', beyond that every finding – the paucity of genes, their 'multi-tasking', those shared 'master genes' – is a nail in the coffin of Darwin's proposed mechanism of natural selection acting on numerous small, random genetic mutations.

From all this, it is possible to glimpse how the Double Helix fails the test of scientific knowability not just once, but twice over. The imperative to condense within those double strands of C G A T molecules the diversity, form and attributes that so readily distinguish man from fly and from all other forms of life would seem to pose an impenetrable barrier to scientific understanding.

The Double Helix fails the further test of scientific knowability because, like Newton's gravitational force, it imposes the order of 'form' on life without there being any evidence of some scientifically measurable objective means by which it might do so. And it fails the test on a much grander scale. It is one thing to try to imagine how the sun might exert its influence on the earth across ninety-two millions of miles of empty space, quite another to try to imagine how that monotonous string of chemical genes can give rise to the millions of species of insect, fish, bird and mammal with which we share this planet. This compels us to recognise, as Baron Cuvier and his fellow natural historians of the nineteenth century inferred, that there must be some non-material formative influence that, from the moment of conception, imposes the order of form on the developing embryo of, say, octopus, squid or kangaroo and holds it constant while its cells and tissues are continually renewed as it grows into adulthood. We cannot, by definition, know that formative 'life force' directly, only infer its reality as the missing factor that might bridge the unbridgeable gap between the 'first order' reality of the phenomena of life as we know it to be in all its wondrous beauty and diversity and the 'second order' reality of its explanation as revealed by those genome projects.

The radical implications of that formative 'life force' will be returned to – but it is appropriate to close this most extraordinary episode in the history of biology, with its 'totally unexpected findings', with two final observations. First, it is astonishing the difference it makes to our everyday lives to acknowledge that we no longer know, as until recently we thought we might, the secrets of genetic inheritance. So while, as ever, a snowdrop is as it is because it is made that way by its snowdrop genes, now, when contemplating a dozen or so of them thrusting their way up through the snow, we are forced to recognise that science cannot tell us how those genes fashion those delicate drooping heads with their evocative white and green colouring. Suddenly it begins to seem as if those snowdrops, and indeed the whole glorious panoply of nature, are infused with a deep sense of the mystery of 'How can these things be?'

And we find too the Human Genome Project ('the most important and significant project that humankind has ever mounted'), for all its scientific sophistication and unexpected findings, paradoxically ends up telling us nothing of real interest about 'humanity'. It tells us nothing, for example, about the skill and courage of the twenty-four-year-old lone yachtswoman Ellen MacArthur who, the same week the 'wondrous map' of the genome was published in 2001, completed her solo circumnavigation of the globe. Her genome, ransacked from top to bottom, would not give the slightest hint of the courage and determination that would have her navigating her boat through the heavy seas of the Pacific Ocean, or clambering to the top of the mast to repair its sails and rigging.

We turn now to the third of those forces of order, the human mind, imposing the order of understanding on the world, which, we can readily anticipate, will similarly fail the test of scientific knowability – with yet more far-reaching implications. For the past 150 years the prevailing view has held the human mind to be 'nothing but' the consequence of the electrical activity of its brain. This, we will find, is not the case.

7

The Fall of Man: A Tragedy in Two Acts

> 'There is no fundamental difference between Man and the higher animals in their mental faculties. It is only our natural prejudices and the arrogance [of] our forefathers which lead us to doubt this conclusion.'
>
> *Charles Darwin*, The Descent of Man (*1871*)

The strangest thing about the universe of which we are part is that there should be 'something rather than nothing'. The second strangest, that we humans are the only beings (as far as we can tell) to know both that it does exist, and the extraordinary events that brought it into being. Without that human presence, 'This moving and sublime spectacle of nature would be sad and mute,' observed the eighteenth-century French philosopher Denis Diderot. 'Everything would be a vast solitude, a phenomenon taking place obscurely, unobserved.'

This uniquely human capacity to know the universe exists is predicated on three separate events, each by itself almost as remarkable as the fact of there being 'something rather than nothing'. The first event was the emergence of 'life' 3,500 million years ago, touched on in the preceding chapter, where the simplest of single-celled organisms is both an astonishingly complex chemical factory and an encyclopaedia of genetic information transmitting the necessary instructions for those chemical reactions from one generation to the next. The second event was the arrival of the earliest marine creatures, such as the trilobite, six hundred million years ago, with the capacity to see, and so be *aware* of the external world – thanks to the twin and simultaneous 'innovations' of both the seeing eye and a brain capable of interpreting the image that falls upon it.

That brain would, over time, become enormously more sophisticated and complex, but the universe would still not become 'knowable' till the third and final event, the emergence of humans, with the faculty of language that permits them to *think*, and then to think about those thoughts and discuss their significance by inserting them into the minds of others gathered around the campfire. Thus, the arrival of our species, witnessed so eloquently by the wondrous art and technology of our earliest ancestors, introduces a radically new element into the universe that had never existed before – the thoughts, values and understanding of what it might all mean. And it does not end there, for that self-reflective human brain comprehends the world from the perspective of the individual to whom it belongs – so that brain must also acquire the sense of that inner person we know ourselves to be, and to whom those thoughts belong. And further, that now reflective self must be free to prefer one thought over another, to argue one interpretation over another and recognise the sovereignty of some 'higher court' than its own immediate impressions: the notion of 'the truth', on which all discourse (whether around the campfire or in the university seminar room) depends – where agreeing with another individual is to acknowledge the *truth* of his argument, and to disagree is to reject it.

There is more, but the gist is clear enough. The privilege of our species in having a larger brain is by itself not sufficient to 'know' the universe. Man must also possess that perception of himself as an 'autonomous self, free to choose' – or, as the eighteenth-century philosopher Adam Smith put it, 'the impartial and well informed spectator' of himself and his thoughts:

> When I endeavour to examine my own conduct . . . I divide myself as it were into two persons; and that I, the examiner and judge, represent a different character from the other I, the person whose conduct is examined into and judged of. The first is the spectator . . . the second is the agent, a person who I properly call myself, and on whose conduct I was endeavouring to form some opinion.

That sense of the autonomous self is more than just a property of the non-material mind, but has a distinctive character whose beliefs and attitudes may change over time but whose personality remains resolutely the same.

> 'That little boy with the shock of black hair that I see in the fading photograph was *me*, now a balding academic in later middle age,' writes the physicist and Master of Queens College, Cambridge, John Polkinghorne. 'The child who was so good at arithmetic but had some trouble in learning to read was *me*, who am now a scientist given to writing about science. Externally, that seems testified to by the history linking that youngster with the current Master of Queen's. Internally, it is testified to by its being my present memory that recalls those early school successes and problems.'

That self, the 'inner person', seems to reside just above and between the eyes looking out on the external world, and is composed of several distinct attributes or parts which, being 'subjective' and 'private', are exclusive to and define their possessor. The first of those 'parts' is the *unique subjective experience* of the world 'out there', that integrates the ineffable quality of the greenness of the leaves of that tree swaying in the wind and the accompanying birdsong into a coherent whole, that is the self's alone. Next, the self is an *autonomous agent*, free to choose, prompted, on its own initiative, to act on the world, to make decisions, to move in one direction rather than another. The third component of the self is the rich autobiographical inner landscape of *memory*, that wealth of accumulated subjective experiences of people, places and events stretching back to childhood. The fourth component is the 'higher' attributes of the mind, those powers of *reason and imagination* that through the power of language transcend the boundaries of personal experience to commune with the minds of others and make sense of the world we inhabit. These several distinct attributes are in turn closely interdependent – so, my subjective impressions of those trees outside my window are influenced by my memories and emotional feelings about trees in general, and so on. Thus, the 'self', though clearly non-material in that it has no substance and cannot be weighed or measured, subdivides into these further interdependent non-material attributes which taken together form the indestructible solid 'inner core' that is each one of us.

That self, to be sure, is grounded in the human brain which gives rise to it, but has its own coherent durable 'reality' that cannot be explained by the ever-changing transient electrical activity of its neuronal circuits. This conundrum of the dual character of a *spiritual* human mind, seemingly floated off from the *physical* brain, has been expressed in several ways. Man is both an 'object', like a table or a chair, the anatomy of whose brain can be objectively described, but also a 'subject', whose understanding of the world derives from his *own* unique private subjective experience. Humanity has both a 'without', an external, visible physical life, and a 'within', an inner, invisible life of the mind.

Man is therefore both part of nature and yet outside it, so while his physical material being is governed by nature and its laws, which are accessible to scientific investigation, the human mind transcends that materiality – and is not so accessible. From Plato onwards the recognition of that durable yet non-material reality of 'the self' with its several component parts crystallised around the concept of *the soul*, which became, within the Christian tradition, the mark of man's exceptionality, which set him apart, guaranteeing him a special relationship with his Creator. The soul was, however, much more than just a theological 'idea', an ethereal substance that departed the body after death. For two thousand years it was perceived as being as durable and real as the physical body itself, the animating principle of man's spiritual being, that unique person by whose distinctive personality others know him to be.

The ascendancy of science in the mid-nineteenth century left no space for the reality of that inner first-person self or soul, whose non-materiality both falls outside and poses a challenge to its exclusive materialist claims to knowledge. That ascendancy rather required that man be knocked from his pedestal and his exceptionality denied, by incorporating him into the evolutionary framework where his mind would become 'nothing but' the consequence of the electrical activity of the neuronal circuits of his brain.

The First Act of the toppling of man from his pedestal starts with the publication in 1871 of Darwin's *The Descent of Man*, which 'crowned the edifice' of his evolutionary theory by fully integrating man within it. The 'Descent' of the title can be a source of confusion, but it refers to 'Descent with Modification' from the common ancestor man shares with his primate cousins, rather than the 'Descent' of being toppled from his pre-eminent position in the Great Order of Things – though clearly the former definition entails the latter. The Second Act opens in the 1960s when evolutionary biologists elaborating on Darwin's arguments in *The Descent* would claim that the full range of unique human attributes – the power of reason and the search for truth; moral sensibility and the distinction between right and wrong; human relationships and the emotions they engender of love and friendship, indifference or antipathy; the virtues of solidarity and compassion – have no inherent value in themselves, but are all genetically inherited traits 'selected for' by nature to ensure 'the survival of the fittest'.

ACT I: THE DESCENT OF MAN

The most persuasive argument for man's evolutionary origin was always the obvious physical similarities (or, as Darwin put it, 'the correspondence in general structure') he shares with his primate cousins, captured most powerfully in Thomas Huxley's famous image of mankind's evolutionary ascent from knuckle-walking primate to upstanding *Homo sapiens*, as considered earlier. Man would, however, remain firmly on his pedestal, and the evolutionary doctrine could not be complete, unless it was also possible to demonstrate that those mental and moral attributes crystallised in the concept of the soul were part of a continuum shared with those primate cousins. Hence Darwin's central argument in *The Descent* is that 'There is no fundamental difference between man and the higher animals in their mental faculties,' so the striking differences between them must be 'of degree but not of kind'. 'It is only our natural prejudices and the arrogance [of] our forefathers which lead us to doubt this conclusion.'

Now, it may be that such is indeed the case, but it is only sensible to scrutinise Darwin's arguments as they unfold in two stages. He first demonstrates that continuum between the minds of animals and man by drawing attention to the many similarities they share. Then he accounts for man's higher attributes, particularly the powers of reason and moral sensibility, by suggesting that they are mental properties of the brain that have evolved to maximise his chances of survival – as demonstrated by the progress of the human species from the 'savage state' to the standards of the 'civilised' races. So, to begin.

Humans and animals, Darwin argued, have much in common. They both feel 'pleasure and pain', 'happiness and misery'; animals 'are exercised by the same emotions' – of terror, suspicion, courage, timidity; they have distinct character traits ('some dogs are good tempered, others turn sulky'); 'they not only love but have the desire to be loved', are 'jealous and feel shame'; 'they feel wonder and exhibit curiosity' and 'have excellent memories for people and places'; they even 'possess some power of imagination', for their movements when asleep suggest they have vivid dreams.

Darwin concedes that man's powers of self-reflection and language pose more of a difficulty, but not insuperably so:

> It may be freely admitted that no animal is self conscious, if by this term it is implied that he reflects on such points as whence he comes or whither he will go, or what is life and death . . . but how can we be so sure that an old dog with an excellent memory and power of imagination never reflects on his past pleasures or pains in the chase? And this will be a form of self consciousness.

The seemingly insurmountable barrier of language, as touched on in an earlier chapter, is resolved in a similar style: 'I cannot doubt that language owes its origin to the imitation and modification of various natural sounds . . . and man's own instinctive cries, aided by signs and gestures.' Thus we learn of a Paraguayan monkey that 'when excited utters at least six distinct sounds, which excite in other monkeys similar emotions', and of a pet African parrot which 'invariably called certain persons of the house, as well as visitors, by their names. He said "Good morning" to everyone at breakfast and "Goodnight" to each as they left the room at night and never reversed these salutations.'

Darwin's argument, with his anecdotal illustrations of the intelligence of dogs, monkeys and parrots, reflects the many remarkable mental attributes of our fellow animals. It can, however, scarcely be described as scientifically rigorous. On the contrary, his is an entirely circular argument that reads the human feelings and emotions of wonder, curiosity, love and jealousy into the behaviour of animals and then infers that they represent a continuum. This would seem unwarranted, as it is impossible to know whether a dog's 'love' or 'curiosity' is the same as our own. It is certainly hard to imagine that Darwin believed his dog's undoubted affection was qualitatively similar to his own devotion to his wife and children; or that its curiosity when out on a walk pursuing the smells of the countryside was similar 'in degree but not in kind' to the curiosity, say, of the astronomers of ancient Egypt who constructed their observatories the better to scrutinise the movements of the planets in the heavens.

The 'higher' functions of the human mind, its powers of reason and moral sensibility, required a different approach. The imperative to incorporate man's moral sense, his 'conscience', into that evolutionary framework was particularly pressing, as so many of its manifestations were so clearly incompatible with the evolutionary doctrine of 'the survival of the fittest' as to be almost an argument against it – most notably caring for the sick, regard for the welfare of the poor, chastity (certainly), respect for the rights of others, courage and self-sacrifice. Further, to many, such as the philosopher Immanuel Kant, the human moral sense, with its ability to distinguish between truth and falsehood, right and wrong, was itself a sign of a divinely ordained universe: 'two things fill the mind with wonder and awe – the starry heavens above me and the moral law within me'.

So Darwin opens the fourth chapter of *The Descent* by acknowledging the difficulty of explaining the sense of moral obligation, as it so frequently requires people to act against their own immediate best interests and so compromise their chances in the struggle of the 'survival of the fittest'.

> It is extremely doubtful whether the offspring of those who were most faithful to their comrades would be reared in greater numbers than the children of selfish and treacherous parents. He who was ready to sacrifice his life rather than betray his comrades would often leave no offspring to inherit his noble nature. The bravest men, who were always willing to come to the front in war, and who freely risked their lives for others, would on average perish in larger numbers than others.

Darwin resolves this difficulty by insisting that the moral sense is a consequence of man's sociability and the obligation to abide by certain rules for the good of the group to which he belonged. Natural selection would then 'select' those societies whose individual members had acquired the trait of a superior moral sensibility, which was then passed on to succeeding generations. Over time societies would progress from the primitive upwards to the high standards exhibited by the civilised races.

Darwin illustrates his concept of evolutionary moral progress by contrasting the culture of the 'savage races' ('immoral and licentious') with that of the 'civilised races'. Thus we learn:

> Most savages are utterly indifferent to the sufferings of strangers, or even delight in witnessing them. Some take a horrid pleasure in cruelty to animals, the women and children of the North American Indians aided in torturing their enemies and *humanity is an unknown virtue* . . . the greatest intemperance is no reproach, licentiousness and unnatural crimes prevail to an astounding extent.

Similarly, their aesthetic sense, 'judging by the hideous ornaments and equally hideous music', was much inferior to that of the civilised races, 'while they are incapable of admiring such scenes as the heavens at night or a beautiful landscape'.

This contentious line of argument would have seemed more self-evident to Darwin than it might appear nowadays, as his views were shaped not just by the prevailing ideology that sanctioned the triumph of the civilised societies over the more primitive, but also by his own experiences on that famous voyage on the *Beagle*.

> The astonishment which I felt on first seeing a party of Fuegians [the inhabitants of Tierra del Fuego at the tip of South America] on a wild and broken shore will never be forgotten by me, for the reflection at once rushed to my mind – *such were our ancestors*. These men were absolutely naked and daubed with paint, their long hair was tangled, their mouths frothed with excitement and their expression was wild, startled and distrustful. They possessed hardly any art, and like wild animals lived on what they could catch; they had no government and were merciless to everyone not of their own small tribe.

Darwin added a further scientific gloss to his interpretation of the ascent of man from the 'savage state' of those Fuegians by citing anatomical comparisons of the brain size of different races: 'There exists in man some close relation between the size of the brain and the development of the intellectual faculties,' he wrote, citing Dr J. Barnard Davis, '[who] has proved by many careful measurements that the mean internal capacity of the skull in Europeans is 92.3 cubic inches; in Americans 87.5, in Asiatics 87.1 and in the Australians only 81.9 cubic inches.'

And so, Darwin argued, natural selection would bring about the progressive enlargement of the savage's brain, and with it that progressive increase in moral sensibility, just as surely as it had produced those wonders of perfection in biology such as the eye, or a bird in flight.

> [Thus, while] a high standard of morality gives but a slight or no advantage to each individual man, advancement in the standard of morality will certainly give an immense advantage to one tribe over another. A tribe including many members possessing in a high degree the spirit of patriotism, fidelity, obedience, courage and sympathy, who were always ready to aid one another and to sacrifice themselves for the common good, will be victorious over most other tribes; and this would be natural selection.

By any standard, *The Descent* is a 'shocker'. 'Darwin had no adequate concept of Man,' observes the philosopher John Greene, nor any sympathetic understanding for those 'savage' societies he disparaged, a misapprehension that the founder of modern anthropology Franz Boas would struggle to correct in the account of his experiences with the 'savage' Eskimos.

> After a long and intimate intercourse with the Eskimo, it was with feelings of sorrow and regret that I parted from my Arctic friends. I had seen that they enjoyed life, and a hard life, as we do; that nature is also beautiful to them . . . and although the character of their life is so rude as compared to civilised life, the Eskimo is a man as we are; that his feelings, his virtues and his shortcomings are based in human nature, like ours.

Still, *The Descent* would find a wide audience among many for whom the validity of its scientific arguments was less important than the persuasiveness of its central idea of human evolutionary progress. Certainly, it seemed a bit harsh that human progress required that the strong should (must) prosper and the weak go to the wall, that the savage should give way to the civilised races, but that was how things clearly were in mid-Victorian England:

> 'Life *was* a struggle: every businessman knew it and if he was honest admitted that a certain ruthlessness in securing the margin between success and failure was inevitable,' commented the historian of science Charles Raven. ' "Nature red in tooth and claw" was a nasty fact, but it was no use crying over it. Sentimentalism was all very well [but] a great nation could not afford to be squeamish . . . so the arguments ran.'

But despite its contentious arguments, *The Descent* would prove much more directly influential than *On the Origin of Species*. For the best part of seventy years it would provide a quasi-scientific rationale for a whole raft of social and political policies, sanctioning the indifference of *laissez-faire* economics, the compulsory sterilisation and elimination of the 'less than perfect' and the doctrine of racial and colonial superiority. Why?

Both *The Origin* and *The Descent* sought, in their different ways, to liberate science from theology. But whereas Darwin, in *The Origin*, found it relatively straightforward to throw over the Biblical notion of 'special creation', the theological interpretation of man he challenged in *The Descent* expressed two profound truths that could not so readily be discarded. The first was the concept of the non-material human soul, as just summarised, with all its seemingly improbable ramifications of Heaven and Hell, the Final Trumpet and the Resurrection of the Dead that can test the faith of even the most ardent of those who profess the Christian faith. There obviously could be no place for the soul in a materialist theory of man, and so it had to go. And yet the soul (or some similar concept) is necessary to capture that truth about the 'within' aspect of the human experience, that ineradicable sense of 'self' as a spiritual entity that grows and changes over time.

The second hidden 'truth' in the Christian doctrine is the notion of an absolute God-given moral law, epitomised most obviously by Moses descending from the mountain with the Ten Commandments chiselled (allegedly) by divine hand on a couple of slabs of rock. There is however more to morality than the Ten Commandments, and the philosophical view would be that the 'moral law within', the distinction between good and evil, the recognition of duty and responsibility, does have an *absolute* quality (where a duty can only be set aside when it ceases to be a duty) and requires humans to apply their *reasoned* judgement to moral dilemmas to find the right course of action. This interpretation of man as a 'free moral agent', capable of distinguishing between right and wrong, clearly sets him apart from his primate cousins, and again can have no place in an exclusively materialist view of man. Rather, as noted, Darwin suggests that moral sense to be 'nothing other' than an inherited trait that encourages social beings to live in harmony with each other, and that can evolve over time to transform the licentiousness of the savage into the high moral standards of a Victorian gentleman scientist.

It would seem hazardous to thoughtlessly set aside the two traditional beliefs in 'the soul' and 'the moral law within' in favour of the contentious judgements, inherent in the principle of natural selection, that favour the strong over the weak and assert the *right* of the strong to prevail. Further, the evolutionary doctrine portrays human suffering not as an evil to be fought against, but as an indubitable good, for 'the war of nations, famine and death', as Darwin expressed it in *The Origin*, are the agents of evolutionary progress. In a speech commemorating the hundredth anniversary of Darwin's birth in 1909, Max Gruber of the University of Munich expressed this most clearly:

> The never-ceasing struggle is not useless. It constantly clears away the malformed, the weak, and the inferior among the generations and thus secures the future for the fit. Thus only through the inexorable extermination of the negative variants does it provide living space for the strong and its strong offspring, and it keeps the species healthy, strong and able to live.

This heady brew of a progressive scientific ideology, untempered by compassion for the vulnerable, and which exalted nature's impersonal laws as 'the highest good', would have the most serious of consequences.

Darwin's evolutionary theory when applied to human society was, at its most benign, no more than an endorsement of the values of the capitalist *laissez-faire* economy of Victorian Britain, while discouraging the well-intentioned efforts of social reformers who might seek to mitigate 'the poverty of the incapable, the distress of the imprudent and the starvation of the idle'. Nonetheless, the inescapable problem remained, as he pointed out, that 'civilised societies' in making their arrangements for the care of the sick and improvident compromised that process of natural selection that had brought them to their preeminent position.

> With savages the weak in body or mind are soon eliminated; and those that survive commonly exhibit a vigorous state of health. We civilised men, on the other hand, do our utmost to check the process of elimination; we build asylums for the imbecile, the maimed, and the sick; we institute poor laws; and our medical men exert their utmost skill to

> save the life of every one to the last moment... Thus the weak members of civilised societies propagate their kind. No one who has attended to the breeding of domestic animals will doubt that this must be highly injurious to the race of man.

Nonetheless, Darwin doubted that there was anything that could be done to prevent these 'highly injurious' consequences. It was, after all, 'the noblest part of man's nature' to feel sympathy for the helpless. Therefore 'we must bear the undoubtedly bad effects of the weak surviving and propagating their kind', and just hope they might refrain from doing so. There were others, however, who would take a more robust view, including Darwin's cousin Francis Galton, for whom Britain's position as the most powerful nation on earth was threatened by the 'deteriorating quality' of its human stock, brought about by the tendency inherent in any civilisation to 'check the fertility of the abler classes... [while] the impoverished and unambitious chiefly keep up the breed'. And so it is that 'the Great Race gradually deteriorates, becoming in each successive generation less fit for high civilisation'. This was the price paid for 'diminishing the rigour of the law of natural selection. It preserves weakly lives that would have perished in more barbarous lands.' The only way out of this dilemma, Galton urged, was to apply the scientific principles of breeding to the betterment of mankind:

> If a twentieth part of the cost and pains were spent in measures for the improvement of the human race as is spent on the improvement of the breed of horses and cattle, what a galaxy of genius might we not create! You might introduce prophets and high priests of civilisation into the world as surely as we can propagate idiots by mating cretins.

Accordingly, Galton founded the science of 'eugenics', or 'good breeding', dedicated to 'the study of those agencies that may improve or repair the racial qualities of future generations'. In Britain this new science attracted widespread support, but never progressed beyond plans to 'encourage the elite drawn from mostly the professional middle classes to have large families', together with advocating 'stern compulsion to prevent the free propagation of those who are seriously afflicted by lunacy, feeble mindedness, habitual criminality and pauperism'.

But elsewhere it was a different matter. The United States, Canada, Sweden and Germany would all legislate in favour of compulsory sterilisation of the physically and mentally disabled by vasectomy in men and tubal ligation in women.

> In Virginia, the state sterilisation authorities raided whole families of 'misfit' mountaineers. The proprietor of a small candy store that catered to those families, Howard Hale, recalled how 'everybody who was drawing welfare was scared they were going to have it done on them . . . they were hiding all through these mountains . . . The sheriff went up there and loaded all of them in a couple of cars and ran them down to the hospital so they could sterilise them.'

In Germany, Darwin's most ardent supporter, Ernst Haeckel, Professor of Zoology at the University of Jena, urged a yet more vigorous application of eugenic policies – whose eventual consequences are only too tragically familiar.

> 'What profit does humanity derive from the thousands of cripples who are born each year, from the deaf and dumb, from cretins, from those with endurable hereditary defects?' Haeckel asked rhetorically. 'What an immense aggregate of suffering and pain these depressing figures represent for the unfortunate people themselves, what a fathomless sum of worry and grief for their families, what a loss in terms of private resources and costs to the state of the healthy.'

Haeckel's proposal that 'a small dose of morphine or cyanide would free the physically and mentally handicapped infant from a long and painful existence' would thirty years later become the official euthanasia policy of the German state, catching in its murderous embrace seventy thousand people whose lives were supposedly 'not worth living'.

Paralleling the new science of eugenics, *The Descent* provided the rationale for scientific racism. Darwin scarcely invented racism, and was himself passionately opposed to slavery, though he had no qualms about describing the human races in terms of 'higher' and 'lower' and likening the 'lower' to the 'higher' apes. But *The Descent*, with its claims about the supposed inferiority of the 'savage', cloaked racism in the aura of scientific respectability, challenging the progressive spirit of the times that they might aspire to a better standard of life. How could they hope to do so when it would require thousands of years of evolutionary progress to acquire the necessary increased brain capacity and moral standards of the European races?

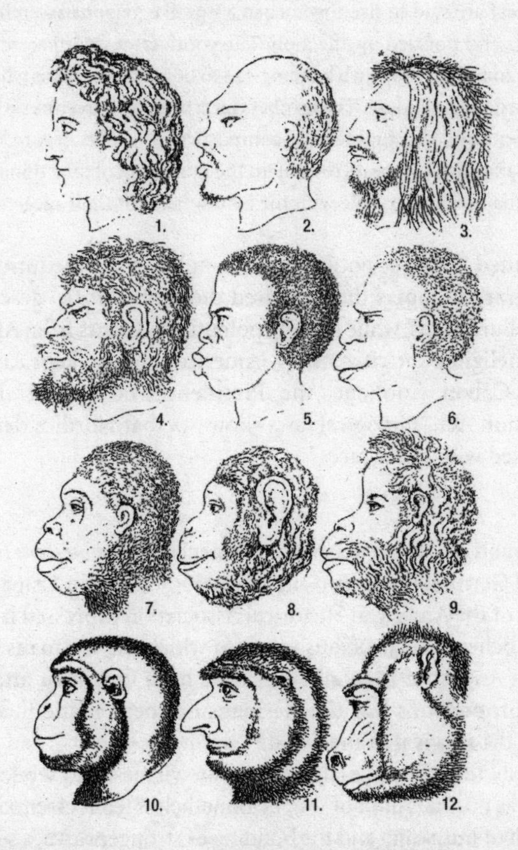

Fig 7-1: *The frontispiece of the first edition of Ernst Haeckel's* The Natural History of Creation *features twelve facial profiles. The first six, purporting to represent the spectrum of human races, start with that of a European, then 'descend' to a Tasmanian. The seventh, a gorilla, is followed by a further five primate species. Haeckel claimed the physiognomic difference between the European and the Tasmanian to be greater than that between the Tasmanian and the gorilla – evidence of man's direct progressive evolutionary ascent from his primate cousins via 'the savage state' to the civilised races.*

The Descent prompted the sort of racial stereotyping epitomised by Francis Galton's description of American Indians, which anticipated their fate in the face of a superior 'civilisation'.

> The men are naturally cold, melancholic, patient and taciturn. [Families] are said to live together in a hut like persons assembled by accident and not tied by affection. The youth treat their parents with neglect and often with such harshness as to horrify Europeans who have witnessed their conduct. The mothers have been seen to commit infanticide without the slightest discomposure . . . The nature of the American Indians appears to contain the minimum of affectionate and social qualities compatible with the continuance of their race.

In the United States, scientific racism would lead to the introduction of immigration quotas that favoured those of 'Nordic' descent over southern Europeans, while barring potential migrants from Africa and Asia as 'ineligible for citizenship'. 'America must be kept American,' declared Calvin Coolidge, the President who signed the 1924 Immigration Act. 'Biological laws show . . . that Nordics deteriorate when mixed with other races.'

The third and more speculative consequence of *The Descent* links it to the rise of German militarism and 'the bloodiest war in history'. As the president of the American Historical Association expressed it in 1918: 'I do not believe the atrocious war into which the Germans plunged Europe in August 1914 would ever have been fought or attained its gigantic proportions had the Germans not been made mad by the theory of the survival of the fittest.'

Certainly that theory exalts militaristic virtues, and while Darwin himself was no champion of war, he nonetheless felt it essential for 'the real health of humanity and the building of stronger races', a sentiment echoed by German military leaders. Vernon Kellog, Professor of Entomology at Stanford University, posted to the headquarters of the German general staff in the early part of the war, recorded the 'disheartening kind of argument' he heard nightly around the dinner table.

> The creed of the Allmacht [almightiness or omnipotence] of natural selection based on violent and competitive struggle is the gospel of

> German intellectuals: all else is illusion and anathema . . . the struggle not only must go on, for that is the natural law, but it should go on so that this natural law may work out in its cruel, inevitable way the salvation of the human species.

Historians of science have understandably drawn a veil over this threefold unhappy legacy of Darwin's evolutionary theory, either glossing over it or portraying its abuses as an aberration – like some powerful weapon that regrettably just happens to have fallen into the wrong hands. That legacy is, however, no aberration, but is an almost inevitable consequence of *any* solely materialist doctrine.

Darwin's evolutionary theory, being objective and scientific, has only room for man as *object*, and so necessarily reduces that 'spiritual' self to the workings of the physical structure of the brain. And that is a very dangerous thing, for to deny the 'within' of human experience, to see humans solely as objects, is to erase what they are, and change them into something else, as *depersonalised* beings. This 'objectification' of man is, as the philosopher Roger Scruton points out, a pervasive feature of materialist and totalitarian systems:

> In such a system, human life is driven underground, and the ideas of freedom and responsibility – ideas without which our picture of man as a moral subject disintegrates entirely, have no public recognition, and no place in the administrative process. It is so easy to destroy people in such a system because human life enters the public world already destroyed, appearing only as an object among others, to be dealt with by experts in the science of man.

Darwin's reactionary (certainly to modern eyes) views cannot be excused as merely reflecting the prevailing views of mid-nineteenth-century Britain. They 'come with the territory' of his evolutionary theory, being endorsed almost universally by most leading biologists in Britain and America in the first part of the twentieth century.

These included Sir Ronald Fisher, who, as described earlier, salvaged Darwin's evolutionary theory in the 1930s by providing that vital, if obscure, mathematical 'proof' of how random genetic mutation might, after all, give rise to the the gradualist evolutionary transformation of one species into another. For Fisher, the doctrine of eugenics promised a 'new phase' of evolution: those of higher ability would, by marrying better and having more children than others, 'become fitted to spread abroad the doctrine of a new, natural nobility of worth and birth'.

They also included Julian Huxley, grandson of Darwin's great supporter Thomas Huxley, who fashioned out of Fisher's obscure mathematics 'the New Synthesis' of Darwin's theory that would revive its fortunes, and who anticipated that 'once the full implications of evolutionary biology are grasped, eugenics will become part of the religion of the future'. In the United States they included Hermann Muller, one of the leading geneticists of the twentieth century, who along with Huxley drew up a 'manifesto' calling for the 'conscious guidance of selection'. And Joshua Lederberg, Director of Genetics at Stanford University, who in 1963 observed: 'The facts of human reproduction are all gloomy – the stratification of fecundity by economic status [the poor and feckless having too many children], and the sheltering by humanitarian medicine of once lethal defects being of particular concern.' And many, many more, not least the co-discoverer of the Double Helix, Francis Crick, who in 1970 urged that 'sterilisation is the only answer to prevent the poorly endowed from having large numbers of unnecessary children'.

In the post-war years the revelations of the practical consequences of 'positive eugenics' in Hitler's Germany would discourage biologists from expressing these eugenic beliefs as frankly and publicly as previously – but the dilemma posed by Darwin's evolutionary theory remains, where the denial of the unique spiritual quality of the 'self' or soul leads inevitably to a misleading interpretation of the human experience.

ACT 2: THE HUMILIATION OF MAN

The conundrum of how to locate man, and his unique mental attributes, within the broad evolutionary framework resurfaced in the 1970s. Biologists may have drawn a veil over Darwin's contentious interpretation in *The Descent*, with all its regrettable consequences, only to substitute a similarly objective and depersonalised explanation where those 'higher' attributes of reason and morality had no value in themselves, but were merely genetically determined traits compelling humans to behave in a way that would maximise their success in propagating their genes.

This interpretation was, in turn, sustained by one of the most astonishing claims in the history of science – that we are not, as we appear to ourselves to be, free and autonomous agents, but are rather the playthings of our 'selfish' genes. We are, apparently, machines created by genes for their own self-propagation – like some throwaway envelope, which they inhabit temporarily for a lifetime, before moving on to the next generation.

> 'They [the genes] swarm in huge colonies safe inside gigantic lumbering robots [ourselves] sealed off from the outside world, communicating with it by tortuous indirect routes, manipulating it by remote control,' writes evolutionary biologist Richard Dawkins. 'They are in you and in me; they created us, body and mind; and their preservation is the ultimate rationale for our existence. We are their survival machines.'

Most people might reasonably suppose this to be some sort of playful joke, perhaps an *ad absurdum* argument to expose the folly of an exclusively materialistic view of man. But it is not, and nor is it just Professor Dawkins – for this represents mainstream conventional evolutionary thinking, taught in schools and universities, expounded in textbooks and popular science, the focus of numerous academic papers every year.

This most unusual (and provocative) take on the human experience begins with the publication in 1975 of a substantial six-hundred-page work, *Sociobiology: The New Synthesis*, by Harvard academic and world authority on ants, Professor E.O. 'Ed' Wilson. His central thesis, that 'Human behaviour is a circuitous technique by which [the] genetic material is kept intact and propagated,' is based on two suppositions. First, he argues, surprising as it might seem, that religion, ethics, family relationships and the many other forms of human culture are determined by the action of specific (if undefined) genes. And second, those genes have been 'selected' in the evolutionary process because they maximise the chance that those who possess them will have a large number of offspring – or, in technical terms, will maximise their 'reproductive fitness'. The central thread that holds Wilson's argument together was the discovery in the preceding decade by evolutionary biologists William Hamilton and Robert Trivers that the altruism and selflessness that is so pervasive a feature of human culture and behaviour is nonetheless compatible with the competitive self-interest that is the driving force of Darwin's evolutionary theory. The argument goes as follows.

There is, and always has been, a clear inconsistency between Darwin's requirement that there be a constant and ruthless struggle between individuals such that only 'the fittest survive', and the everyday reality of human society, which is marked by the altruism and mutual support necessary for any sort of cooperative endeavour. Hence (it is supposed), a long time ago, when humans lived 'in a savage stage', their life was, as Thomas Huxley put it, 'a continual free fight where each man appropriated whatever took his fancy and killed whatever opposed him'. Humans eventually emerged from this savage state to develop the forms of attachment and loyalty, epitomised by marriage, religion, laws and customs, that are a universal feature of human society. It is not clear precisely when this turnaround took place, but it must have been several million years ago, for no other reason than that the human infant is quite uniquely helpless, and thus some form of mutual cooperation is essential simply to rear the next generation.

So clearly there is a powerful incentive for humans to cooperate, to be altruistic towards each other; but as the Australian philosopher David Stove has pointed out, this does not stop altruism causing a most serious 'difficulty':

> Altruism ought to be non-existent, or short-lived wherever it does occur – if the Darwinian theory of evolution is true. By the very meaning of the word, altruism is an attribute which supposes its possessor to put the interests of others before his own. Disposes him, for example, to defend [his family] in danger, when he could have simply saved his own skin; disposes him to eat less, or less well, if this helps others to eat more, or better. But any such behaviour clearly tends to lessen his own chances of surviving and reproducing; and altruism is therefore an attribute which is injurious to its possessor in the 'struggle for life'.

Darwin had dealt with the 'difficulty' of human altruism, the tendency to put the interests of others above one's own, in the context of that supposed trajectory from the 'savage' to 'civilised' races, claiming that nature would 'select for a higher standard of morality', as it would improve the survival prospects of the tribe that possessed it over another that did not. This interpretation might, for obvious reasons, be no longer acceptable, but the problem of reconciling altruism with the 'survival of the fittest' persisted until 1964, when an English biologist, William Hamilton, proposed it was soluble after all – just so long as everyone realised that the purpose of natural selection was not (as most might suppose) to maximise the survival of the 'fittest' *individuals*, but the successful propagation of those genes shared with one's immediate relatives. To clarify (and clarification is certainly needed).

While an undergraduate at Cambridge University in the 1950s, William 'Bill' Hamilton was much impressed by Sir Ronald Fisher's *Genetical Theory of Natural Selection*, whose obscure mathematical proof of natural selection and eugenic beliefs (which he shared) have just been touched on. Hamilton was drawn to the persistent problem of the human propensity to altruism by reflecting on that remarkable biological phenomenon where many species of insects – ants, bees and wasps – possess a 'caste' of sterile or infertile 'workers' who spend their lives looking after their mother, the 'Queen', and caring for their siblings – but have no offspring themselves. The lives of these sterile workers are purely 'altruistic', as they are not promoting their own self-interest, but that of their immediate relatives. So, while the sterile ant would perish with no progeny and her own genes would die with her, those genes she shared with her immediate relatives would, thanks in part to her efforts, live on. This led Hamilton to formulate the concept of 'inclusive fitness', where the purpose of natural selection is expanded to include promoting the best interests of the entire ant community. There must, in short, be genes for 'altruism' that direct the sterile ants in their concern for others, and that are preserved within the gene pool of the species through the greater good they confer *to the species as a whole.*

Hamilton found a way to express this idea in mathematical form, by showing that the closer the relationship of one person to another, the more powerful the action of this 'altruistic' gene. So, father and mother, both with a major 'investment' (50 per cent each) in their children's genes, are more inclined to care for them, even to their own personal disadvantage, feeding them, nursing them, playing with them, than they would be towards a nephew or niece, with whom they only share a quarter of their genes. Or, to be specific, '[While] no one is prepared to sacrifice his life for any single person [they will do so] *for two or more offspring* [each of whom shares 50 per cent of their parental genes, making up 100 per cent in total], *or four half-brothers* [by the same calculation, 4 x 25 per cent], *or eight first cousins* [again by the same calculation, 8 x 12.5 per cent]'.

Hamilton's mathematical formula, as with Fisher's 'proof' of natural selection cited in an earlier chapter, is included here not for the purpose of clarification – but to convey its similarly salient characteristic of incomprehensibility.

Every effect on reproduction which is due to A can be thought of as made up of two parts: an effect on the reproduction of genes i.b.d. with genes in A, and an effect on the reproduction of unrelated genes. Since the coefficient r measures the expected fraction of genes i.b.d. in a relative, for any particular degree of relationship this breakdown may be written quantitatively:

$$(\delta a_{\text{rel.}})_{\text{A}} = r(\delta a_{\text{rel.}})_{\text{A}} + (1-r)(\delta a_{\text{rel.}})_{\text{A}}.$$

The total of effects on reproduction which are due to A may be treated similarly:

$$\sum_{\text{rel.}}(\delta a_{\text{rel.}})_{\text{A}} = \sum_{\text{rel.}} r(\delta a_{\text{rel.}})_{\text{A}} + \sum_{\text{rel.}}(1-r)(\delta a_{\text{rel.}})_{\text{A}},$$

or

$$\sum_{r}(\delta a_{r})_{\text{A}} = \sum_{r} r(\delta a_{r})_{\text{A}} + \sum_{r}(1-r)(\delta a_{r})_{\text{A}},$$

which we rewrite briefly as

$$\delta T^{\bullet}_{\text{A}} = \delta R^{\bullet}_{\text{A}} + \delta S_{\text{A}},$$

Fig 7-2: *William Hamilton's bafflingly complex mathematical proof that parental affection is mainly determined by genetic self-interest.*

The reasonable suspicion that this is mathematical obscurantism will become clearer after briefly considering the further discovery, seven years later, by Robert Trivers of Harvard University, that solved the yet more substantial aspect of the 'problem of altruism' – why humans should dedicate their lives to caring not just for their immediate kind, but also for the poor and needy to whom they are completely unrelated. Why, for example, should Albert Schweitzer have felt impelled to found a leprosy hospital in the inhospitable forests of central Africa? Trivers's explanation, published in a series of academic papers from 1971 onwards, lay in 'reciprocal altruism' – people might *appear* to make altruistic sacrifices on behalf of others, but in reality they are directed to do so by their genes in anticipation that others will reciprocate.

Trivers, by way of illustration, reflects on the calculations that might lead someone to save a man from drowning:

> Assume that the chance of the man drowning is one in fifty – if no one leaps in to save him. [Assume too that] the chance that his potential rescuer will drown if he leaps in to save him is much smaller, say one in twenty. Assume that the drowning man always drowns when his rescuer does and that he is always saved when the rescuer survives the rescue attempt. Were this an isolated event, it is clear that the rescuer should not bother to save the drowning man [because he runs the risk of drowning]. But if the drowning man reciprocates at some future time, *it will have been to the benefit of each participant to have risked his life for the other.* Each participant will have traded a one half chance for about a *one tenth* chance of their dying. If we assume that the entire population is sooner or later exposed to the same risk of drowning, the two individuals who risked their lives to save each other will be selected over those who face drowning on their own.

Trivers's mathematical reasoning of the 'net' benefit accruing to someone behaving altruistically by this reasoning is cited here for the same reason as earlier:

What is required is that the net benefit accruing to a typical a_2a_2 altruist exceed that accruing to an a_1a_1 non-altruist, or that

$$(1/p^2)(\Sigma b_k - \Sigma c_j) > (1/q^2)\Sigma b_m,$$

where b_k is the benefit to the a_2a_2 altruist of the *k*th altruistic act performed toward him, where c_j is the cost of the *j*th altruistic act by the a_2a_2 altruist, where b_m is the benefit of the *m*th altruistic act to the a_1a_1 nonaltruist, and where p is the frequency in the population of the a_2 allele and q that of the a_1 allele..

Fig 7-3: *The opening premise of Robert Trivers's 'The Evolution of Reciprocal Altruism', published in the* Quarterly Review of Biology, *1971.*

Here then, Trivers claimed, is the explanation for the full complex of human emotions, of friendship, gratitude, sympathy, trust and so on. They are all genetically inherited traits 'selected by nature' in order to ensure the reciprocal altruism that will promote the best interests of their possessor – and thus the propagation of their genes. We feel gratitude towards those who are kind to us because the gratitude gene 'has been selected as being sensitive to the costs/benefit ratio of altruistic acts' – a claim that raises several substantial questions. Where, one might ask, are these genes for altruism, friendship and generosity? How is one meant to suppose that humans (let alone animals) might make the necessary mathematical calculations before committing themselves to acting altruistically towards others? How do people keep the 'tit-for-tat' tally that would encourage them in their altruistic behaviour, in anticipation that the recipient will repay his dues by returning the compliment at some time in the future?

It can admittedly be very difficult to comprehend the mindset of anyone who might believe that the noble virtues of parental love and human compassion are nothing other than the self-interested calculation of their genes. But the significance of Hamilton's and Trivers's mathematical formulae, like Fisher's 'proof' of natural selection forty years earlier, lay in conferring an aura of scientific objectivity on the 'solution' of that most obvious anomaly of the evolutionary theory – how to reconcile the 'struggle for survival' with the perverse tendency of humans to prioritise the interests of others over their own. Mathematics apart, the notion that love, sympathy and compassion are 'nothing other' than a deception foisted by selfish genes on the human mind would prove very influential.

> 'The economy of nature is competitive from beginning to end,' observes evolutionary biologist Michael Ghiselin. 'No hint of genuine charity ameliorates our vision ... the impulses that lead one animal to sacrifice himself for another turn out to have the ultimate rationale in gaining advantage over a third. Scratch an "altruist" and watch a "hypocrite" bleed.'

This contrarian interpretation of human altruism as covert genetic self-interest was plucked from (well-deserved) obscurity by Professor Ed Wilson, who recognised its wider implications: the entire spectrum of human activity and behaviour could now be accounted for as nothing but 'a circuitous technique for keeping the genetic material intact'. Compassion, we learn, is 'ultimately self-serving'; that is to say, it conforms to the best interests of self, family and allies of the moment. 'The [compassionate person] expects reciprocation from society, for himself or his closest relatives. His good behaviour is calculating and his manoeuvres orchestrated by the intricate sanctions and demands of society.'

Similarly, the purpose of the ennobling human virtues of courage and loyalty is to reinforce the principle of reciprocal altruism ('lives of the most towering heroism are paid out in expectation of great reward'), while the opprobrium attached to the cheat and the traitor is intended to discourage those who might break those rules of reciprocation.

Contrariwise, while many might suppose human aggression to be a reprehensible moral flaw, Professor Wilson suggests it is an evolutionary 'adaptation' by our ancestors to threats to their territory:

> Humans are strongly predisposed to respond with unreasoning hatred to [such] threats ... We tend to fear deeply the actions of strangers and to solve conflict by aggression. These rules are most likely to have evolved during human evolution and to have conferred a biological advantage on those who conform to them with the greatest fidelity.

The near-universal prevalence of religious faith in every human society is similarly explained. 'Human beings', we learn, are 'notoriously easy to indoctrinate', because their religious faith is similarly genetically determined, conferring the biological advantage that comes from supposing that there is a purpose or meaning to life. The motivation of those, such as Albert Schweitzer, who commit their lives in a most non-Darwinian way to the service of others, is readily explicable as being driven by the anticipation of the long-term benefit from their selfless actions of reaping their reward in heaven. Wilson claims, in short, that we are deluded to suppose that we are genuinely moved to compassion by the plight of others, or admire the heroism of those who resisted Nazi tyranny, or infer from the 'sublime' of the world around us the existence of a higher intelligence. Rather, such feelings and inferences are a ploy to promote our self-interest, by making others well-disposed to us or making us feel good about ourselves.

Wilson's sociobiology would in turn spawn a new branch of psychology, 'evolutionary psychology', that applies the same insights to explain away the full range of human feelings and emotions – jealousy, fidelity, status-seeking and many more – as 'modules' in the brain genetically inherited from our Stone Age ancestors, when they were 'selected' to maximise their chances of reproductive success. To take just one example, while we might naïvely suppose conjugal love to be a particularly ennobling characteristic of our species, it turns out to be 'a marvellous piece of evolutionary engineering' to ensure that both parents will care for their children, whose survival will reward their mutual genetic investment in them.

Its purpose is to resolve the (supposed) 'asymmetry of interest' between men and women when it comes to propagating their genes – where, as the writer Gore Vidal puts it, 'Boys are made to squirt and girls are made to lay eggs. And if the truth be known, boys don't really much care what they squirt into.' So, whereas men with their abundant sperm have an obvious interest in having sex with as many women as possible in order to maximise the propagation of their genes, women must be choosy with whom they mate, as their 'investment' comes in the form of just a single fertilised egg, limiting the number of their offspring to a maximum of about one a year throughout their reproductive life.

This 'asymmetry of interest' means that men have evolved to be 'easily aroused', with a 'limitless appetite for casual sexual partners', observes the most prominent of evolutionary psychologists, Professor Steven Pinker of the Massachusetts Institute of Technology: indeed, they have an 'insatiable' desire for 'a variety of sexual partners for the sheer sake of . . . variety'. They will thus seek out 'loose' women to propagate their genes, but thereby run the risk of being 'cuckolded' and ending up being responsible for offspring who are not their own. Therefore they must also, if they are to be as certain as possible of the paternity of their children, enter into a long-term pair-bond arrangement and jealously enforce their partner's fidelity. This double standard turns out to be the 'optimum genetic strategy: mate with any female that will let you but make sure your consort does not mate with any other male'.

Women, on the other hand, exert control over the stronger and dominant males by using 'copulation as a service or favour' (such blunt terminology pervades evolutionary psychology, the better to emphasise the continuity between the sexual behaviour of animals and humans). Meanwhile, their pattern of ovulation has evolved so that its timing is concealed, coercing desirable males into consort relationships, as 'he would have to mate with her over a considerable span of the sexual cycle if he was to make her pregnant with his own offspring'.

From this perspective it is possible to see how conjugal love might have evolved 'to ensure males and females stick together long enough to rear their current offspring'. Or, to put it more bluntly still, 'The man who "sells" his wife a genetic half-interest in his children gets in return more than someone who will take a share in the rearing of the children,' writes Richard Posner of Chicago University. 'He gets a child rearer who has a superior *motivation* to do a good job precisely because of the genetic bond. Altruism is a substitute for market incentives and the man can take advantage of the substitute by giving his wife a genetic stake in his children.'

Some men, of course, are not up to scratch, and without the necessary resources to support a partner are unlikely to find a consort. This explains the motivation of the rapist, who, as a loser, must resort to coercion as the only way to propagate his genes: 'Rapists can detect women who are more likely to carry a child to term and preferably attack these women,' observes evolutionary psychologist Craig Palmer of the University of Colorado and co-author of *The Natural History of Rape*. At this point, one might think that the evolutionary 'science of man' is beyond caricature, and can sink no lower.

What, then, to make of sociobiology and its offspring, evolutionary psychology? Does it 'solve the problem of altruism in human behaviour', or is it, as biologist James C. King of the University of New York would claim, 'a shocking attempt to ensnare us all in a pseudo-scientific set of rules [compounded of] obsolete genetics and a cynical interpretation of social relations'.

Professor King's reference to 'obsolete genetics' takes us to the heart of the matter, for the most striking feature of this modern-day evolutionary science of man is the dismal quality of the science that underpins it – not because one might disagree with its findings or implications, but because it has no recognisable basis in reality. Fictional selfish genes give rise to equally fictional brain 'modules' of jealousy, fidelity and so on, whose fictional (and reactionary) instincts and emotions give rise in turn to a fictional (and debased) portrayal of humanity.

It is difficult enough to try to conceive, as we have seen, how the few thousand genes of the human genome can somehow contain the instructions for so complex a phenomenon as the human brain. But it is another thing entirely to suppose some genes might pursue their own selfish interests (as if that were possible) to 'cause' something as elusive as selflessness, or compassion, or any of those hypothetical evolutionary-determined genetic traits that supposedly explain human behaviour. 'No one has ever been able to relate any aspect of human social behaviour to any particular gene or set of genes,' observes the geneticist Richard Lewontin. 'Thus all statements about the genetic basis of human social traits are purely speculative.'

The failure to identify those non-existent altruistic genes in the human genome has proved no impediment to their being highly influential. Current evolutionary textbooks maintain that Hamilton and Trivers enormously expanded the explanatory power of Darwin's theory of natural selection by reconciling the 'survival of the fittest' with the phenomenon of altruism. Certainly, when pressed, evolutionary biologists might take exception to being themselves the playthings of their selfish genes, compelling the men amongst them to heights of sexual promiscuity and the women to using 'copulation as a service or favour'; but then there is always the let-out clause that we modern humans alone have managed to escape their clutches – which rather undermines the role for which they were invented, as the determinants of human behaviour and culture.

It can be difficult to understand the appeal of sociobiology and evolutionary psychology, other than to note that its central theme has much in common with the popular 'veneer idea', where the virtues of self-sacrifice and consideration for others are portrayed as a thin disguise to conceal our fundamentally selfish and non-moral animal nature.

> 'There is a perennial type to whom this belief is peculiarly and irresistibly congenial,' observes David Stove. 'It is the kind of man who is deficient in generous or even disinterested impulses himself, and knows it, but keeps up his self-esteem by thinking that everyone else is really the same . . . he prides himself on having the perspicacity to realise what most people disguise even from themselves – that everyone is selfish, and having the uncommon candour not to conceal this unpleasant truth.'

It is, as ever, necessary to keep an appropriate sense of proportion in all this. Human selfishness runs very deep. People act for the most part in their own self-interest (it would be absurd not to do so) and, 'blood being thicker than water', seek to safeguard the interests of the family, group or nation to which they are bound by the ties of familiarity and common culture. True human sympathy rarely stretches beyond one's immediate circle – or, as the French philosopher La Rochefoucauld put it, 'We all have strength enough to endure the misfortune of others.' But it is equally self-evident that the contrary holds: compassion, self-sacrifice, courage and solidarity are not only pervasive, but those whose lives are testimony to these and other virtues are held in high regard. Further, that high regard is predicated on the presumption that humans are responsible for their actions, and by acting in this way make their own non-obvious choice of placing the interests of others over their own. It may be that this poses a problem for Professor Ed Wilson and all who subscribe to an evolutionary doctrine centred on the premise of the survival of the fittest, but if they wish to spend their lives chopping logic and devising obscure mathematical proofs that we humans are in reality the playthings of our genes, that is their misfortune.

There is, however, another 'problem' of a different sort, that centres on the near-universal presumption that Darwin, in solving the problem of man's origins, provided a materialistic, non-spiritual explanation for his mind. And the presumption, too, that in doing so he liberated man from the superstitions of the past, to confront the harsh reality that he is 'nothing else and nothing higher' than the fortuitous consequence of a natural process: 'the struggle among individual organisms to promote their own personal reproductive success'.

But that Fall of Man, toppled at last from his pedestal to confront the meaninglessness of his existence, has resulted, as we have seen, first in the most grievous social policies and, second, in him being deprived of his freedom, to become no more than a plaything of his genes. The source of all this mischief lies in the necessity to portray man not *as he is*, but as he *has to be* in order to incorporate him into an evolutionary theory that requires him to be different 'only in degree but not in kind' from his primate cousins.

We need, in short, a fuller, more rounded view that acknowledges the core reality of the human experience which sets us apart – the sense of the autonomous, independent 'self' not as some shadowy, elusive entity, but something real and tangible that explains the force of character and the personality that is within each of us. And where is that 'more rounded' view to come from, but from science itself?

There can, at times, be a pleasing symmetry in events. By the late 1980s, Darwin's evolutionary theory as set out in *The Origin* was in the most serious trouble – but it took science, in the form of the New Genetics, to deliver the fatal blow, with its revelation of the near-equivalence of the genomes of man, mouse, primate, fly and worm. Now, it would seem, Darwin's arguments as they applied to man, set out in *The Descent* and elaborated by Hamilton, Trivers, Wilson and others, are similarly in the most serious trouble, and it is the turn of the neurosciences of the Decade of the Brain to deliver their own verdict on this materialistic explanation of the human mind. And so they do, and in the process stumble almost inadvertently on one of the most important scientific insights of this, or any, century.

8

The Limits of Science 3: The Unfathomable Brain

The brain is wider than the sky,
For, put them side by side,
This one the other will include
With ease, and you beside.

Emily Dickinson

The sun, for all the warmth and energy and beauty with which it suffuses our lives, is a trivial fact when set against the human brain. Certainly its size, 99 per cent of the matter of our solar system, its core temperature of 26,000 degrees Centigrade, and its estimated lifespan of ten thousand million years are impressive enough. But it is no more than a massive exercise in nuclear fusion – converting, every second, seven hundred million tons of hydrogen into 695 million tons of helium and five million tons of heat and energy in the form of gamma rays, which traverse ninety million miles of space to initiate the great cycle of life here on earth. That is pretty much all it does. By contrast, the human brain is so deep and so talented as to defy the most thorough exposition of its attributes. Moment by moment, it perceives the world 'out there' in all its exquisite detail, stores its experience and knowledge to be instantly recalled decades later, comprehends through the powers of reason and imagination the natural world of which it is a part, and realises the imaginative genius of every painter, poet, writer and composer. The intellectual compass of the brain, though confined within the silent darkness of the skull, is, as Emily Dickinson puts it, 'wider than the sky' – both across *time* when reflecting on the past, understanding the present and predicting the future; and across *space*, traversing every order of magnitude from the near-infinite vastness of the universe to its antithesis, the near-infinitesimal smallness of a single atom.

The brain too poses the greatest conundrum within that universe: how the same three pounds of protoplasmic stuff can contain the distinctive character and personality that is each one of us – both the billions of humans with whom we share the planet, and all those who have ever lived. Or, as the German philosopher Friedrich Nietzsche put it, 'Every man knows well enough he is a unique being only once on this earth [that will] never be put together again.' The phenomenon of human individuality is the source too of virtually, if not quite all, that is important. The distinctiveness of human character is the very essence of our social relationships, for the personalities of those we know (or might come to know) – lovers past and present, our parents and children, friends, relations and transient acquaintances – and the feelings we have for them, provide most of the colour, interest and meaning of our existence. Human individuality too is the foundation of human freedom, for we feel ourselves to be free precisely because our thoughts and beliefs are uniquely our own, and distinct from those of others. It is not difficult to imagine how impoverished human existence would be without that distinctiveness – were we, like the clones of science fiction, mere replicas of each other.

The significance of human individuality could be spun out with further examples almost indefinitely, but the central enigma is clear enough: how to reconcile what the brain *is* with what it *does*? How can the electronic firing of those billions of neurons of the same monotonous physical structure of the brain be the entire causal basis of so vast a range of mental life, the near-infinite diversity of our individual selves, with our own unique thoughts, memories and beliefs?

The obvious answer would be that they cannot be reconciled, and the dissonance between the unprepossessing, homogeneous brain and the spiritual mind to which it gives rise was for thousands of years the most persuasive evidence for the 'dual' nature of reality consisting of both a material and a non-material domain.

The founder of modern philosophy, the Frenchman René Descartes in the seventeenth century, was the first to make this distinction between brain and mind with great clarity, pointing out how the essence of the physical material brain is *qualitatively* different from that of the spiritual mind, because whereas physical objects (such as the brain) occupy space, the mind and its thoughts do not. Hence, the methods of science with which we investigate the physical brain *directly*, observing its activity, weighing and measuring, discovering through experiment its mechanical properties, are qualitatively different from the methods of reflection, introspection and 'philosophising' in its broadest sense with which we seek to understand the mind. So too, logically, the forms of knowledge arrived at from these methods of investigation are also necessarily qualitatively different. The facts of the brain are objective and independently verifiable. The facts of the mind – its thoughts, memories and beliefs – are subjective, and only directly knowable to its possessor. They must be causally linked, for, self-evidently, injury to the brain impairs the thoughts and emotions of the mind. Still, the commonsensical interpretation would be that they are two distinct, if related, 'things'.

That interpretation could, as noted, scarcely survive the ascendancy of materialist science in the mid-nineteenth century, with its denial of the most powerful intuitive evidence (or so it seemed) of the non-material domain – the coherent and durable self, or soul, at the centre of the human experience. We examined in the preceding chapter the attempt to explain away those distinctive non-material features of man's mind by incorporating him within the broad framework of a materialist evolutionary theory – and noted its many unfortunate consequences. We now turn to the second parallel approach, the promise that the remorseless onward march of science would collapse the non-material domain into the material, by locating that autonomous self, 'free to choose', and its several interconnected parts – the personal, subjective experience of the world 'out there', the faculties of reason and imagination, memory and experience and so on – in the physical structure of the brain. This posed, admittedly, many substantial difficulties. First, science could offer no theory as to how the physical activity of the brain might translate into the thoughts and perceptions of subjective experience – as Thomas Huxley acknowledged, in a most telling metaphor from a familiar story: 'How it is that anything so remarkable as a state of consciousness [awareness] comes about as a result of irritating nervous tissue [the activity of the brain] is just as unaccountable as the appearance of the Djinn when Aladdin rubbed his lamp.'

The second difficulty is the problem of *mental causation*, or 'free will': how (at its simplest) the non-material thoughts of the mind can influence the workings of the brain, activating the neural circuits that cause us to choose one course of action over another. This facility to make choices of our own 'free will' is rightly amongst the most treasured attributes of the human mind, realised a hundred times a day even in so trivial an action as crossing the road – we *decide* to do so at one moment rather than another. But to accept the supposition that non-material thoughts (the desire to cross the road) can have physical effects (causing the legs to move) would be to introduce into our understanding of the natural world some non-material force that stands outside, and is not governed by, the principles of lawful material causation. This dilemma can only be resolved in materialist terms by supposing that the decision (for example) when to cross the road is not freely taken, but is determined by the electrical activity of our brain. By this count we are not just the 'plaything of our genes' of the preceding chapter, but also the stooge of our brains, and the impression we all have of being 'free to choose' is an illusion generated by our neuronal circuits. Finally, the most difficult of all challenges for an exclusively materialist explanation of the brain is to locate within the electrical activity of its neuronal circuits that sense of the durable self or soul, with its own distinctive personality that changes over time and yet remains the same, presiding over that inner landscape of thought, memory and emotion, looking out on and making sense of the world 'out there'.

In the light of these difficulties, there seemed no alternative other than to acknowledge the prevailing limits of scientific understanding, but nonetheless insist that the human mind would eventually be shown to be reducible to, explicable in terms of, the physical activity of the brain. How could it be otherwise? Descartes' dualism of physical brain and non-material mind as two different 'essences' offered no hint as to how they might be linked together. It seemed much more logical to argue that *somehow* they were one and the same, as Francis Crick, the co-discoverer of the Double Helix, would put it:

> 'You, your joys and your sorrows, your memories and your ambitions, your sense of personal identity and free will, are in fact no more than the behaviour of a vast assembly of nerve cells and their associated molecules.'

The pursuit of that vision of an objective, scientific explanation of the human mind in terms of the physical workings of the brain would come in three instalments, prompted initially by the discovery towards the close of the nineteenth century that the brain was not nearly as mysterious and inexplicable as it appeared: its convoluted cerebral hemispheres proved to be a chequerboard of specialised functions, while its dense gelatinous material, when examined under the microscope, was transformed into a dazzling and intricate tracery of nerve fibres whose interconnections might all too readily account for many of the attributes of the non-material mind.

1. Mapping the Territory: 1861–1950

The external map of the brain, parcelled up into its several provinces, is by now so familiar that it is difficult to imagine how for thousands of years it was as blank as an unexplored continent. Then, in 1861, a French neurosurgeon, Pierre Paul Broca, discovered its first major landmark, the speech centre on the left side of the brain. In this well-known story, Broca admitted under his care at the Hôpital Bicetre in Paris a man known to all as 'Tan' – the only sound he had been able to utter since suffering a stroke thirty years earlier. Tan rapidly succumbed from the gangrene affecting his limbs, and at the subsequent autopsy Broca identified a discrete area of damage in the posterior part of the left frontal lobe, and drew the obvious conclusion: '*Nous parlons avec notre hémisphère gauche*' ('We speak with our left hemisphere'), he declared, precipitating a gold rush of similar findings. Before long the German neurologist Karl Wernicke described a comparable defect in patients who were able to talk fluently but were unable to comprehend speech, and who at autopsy displayed evidence of damage to the posterior part of the temporal lobe – the first indication that distinct components of the same mental function, the faculty of language, were processed in different parts of the brain. The casualties of the trenches in the First World War offered many opportunities to further refine 'the map': most notably, the Irish neurologist Gordon Holmes tested the visual fields of soldiers who had sustained bullet and shrapnel injuries to the visual cortex at the back of the brain, and found that even a tiny area of damage could produce a corresponding patch of blindness.

From the 1930s onwards the most prolific of the brain's cartographers, the Canadian neurosurgeon Wilder Penfield, investigated the effects of stimulating the surface of the brain with a weak electric current while his patients were on the operating table. He was able to locate both sensation and movement to the two discrete bands of the sensory and motor cortex running down both sides of the brain, and showed how the most sensitive parts of the hands, mouth and genitalia were disproportionately represented.

By 1950 the great project to map the properties of mind to specific regions of the brain was, at least in broad outline, essentially complete: the motor and sensory cortex, each representing discrete parts of the body, occupy the two strips down the lateral aspect of either side; the language centre occupies a diffuse part of the left hemisphere; the visual cortex is in the posterior lobe; and the dominant frontal lobes are the site of higher mental functions such as planning and rational thought.

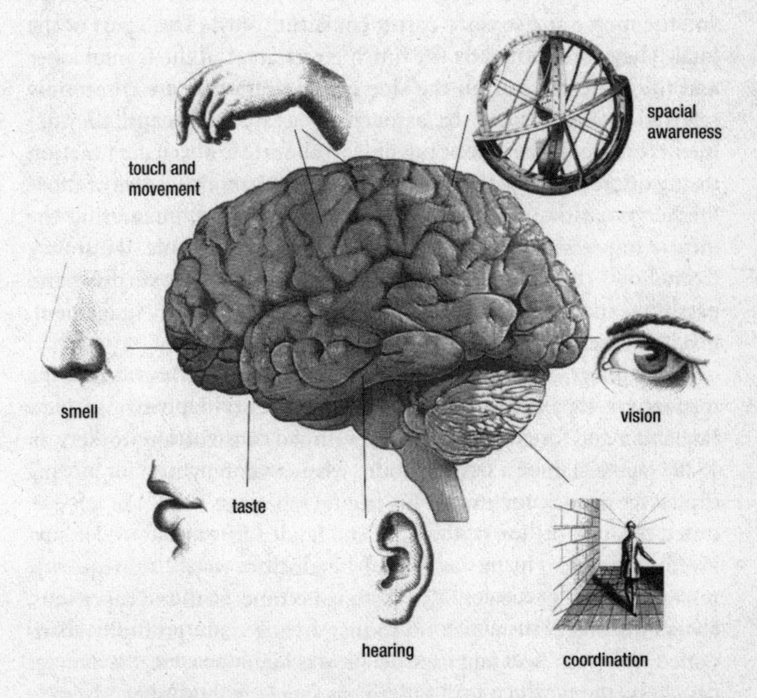

Fig 8-1: *The specialisation of the parts of the brain to fulfil its many different functions includes (as illustrated here) the sensory perception of vision, smell, hearing and taste. The frontal lobes are dedicated to the 'higher' attributes of reason and imagination, and large tracts of the left hemisphere to language.*

The parcelling up of the real estate of the brain to its distinct zones of interest perhaps suggests a deeper understanding of the workings of the brain than is warranted – the allocation of vision to the visual cortex or hearing to the auditory cortex might imply that we know *how* we see or hear the world 'out there'. On the contrary, the structure and electrical activity of these parts of the brain are virtually indistinguishable, and thus these senses' localisation to these discrete parts of the brain offers no insight into how their respective neuronal circuits 'translate' into such qualitatively utterly distinct subjective experiences as watching the sunset or listening to a Bach cantata.

Then, it might seem, when looking at the map of the brain's territories, that most of it is accounted for by one function or another. Again, and on the contrary, those speech centres, the visual cortex and the motor and sensory cortex constitute only a small part of the total. They are dwarfed by the much larger areas of the frontal lobes and the parietal lobes on the side of the brain that are commonly referred to as the 'silent' or 'associative' areas, not because they are 'silent', but because it is not possible to allocate any specific function to any discrete area within them. Rather, they are the centre of those 'higher' cognitive properties of the human mind, integrating the diffuse impressions of the senses into a coherent whole, the source (somehow) of the human emotions of love, hatred, surprise and passion – and of those intellectual attributes of wisdom, judgement, insight and creativity.

These integrative functions of the frontal lobes are illustrated by the misfortune that befell Phineas Gage, a twenty-five-year-old New Englander and foreman of a team of railroad construction workers. In 1848 Gage sustained a terrible injury when a tamping iron for putting explosives in rock penetrated his frontal lobes just below the left eye, exited through the top of the skull and landed fifty feet away. He survived, astonishingly, but as one of his doctors would subsequently remark, he was 'no longer Gage', having become 'fitful and capricious, always making plans which no sooner arranged are promptly abandoned'. Whereas Tan's language deficit was highly specific, the damage caused by the tamping iron to Phineas Gage's frontal lobes adversely affected *all* the higher aspects of his mind – his decision-making, problem-solving, judgement, sympathy, competence and so on.

The endeavours of Pierre Paul Broca, Gordon Holmes, Wilder Penfield and many others in mapping the brain marked the first and essential step in unravelling its secrets, but future progress would require something intellectually more substantial, a way of imagining how the physical brain might give rise to the non-material mind. In 1936 the brilliant British mathematician Alan Turing conceived of a 'universal machine' that could in principle carry out any mathematical task using a 'binary' code of just two symbols, a '1' or a '0'. Similarly, the individual neurons in the brain too have two modes – they can either 'excite' or 'inhibit' the electrical activity of other neurons in close proximity. So the brain acquired a new metaphor – the digital computer – that would define the second instalment of its unravelling, and in particular how the synthesis of 'nature' (the 'hardware' of its neuronal circuits) and 'nurture' (the programming 'software' of experience) might conjure from its monotonous structure the unique individuality of character that is each one of us.

2. The Brain as Computer: 1950–1980

Both computers and brains are prodigiously intelligent, though, set side by side, the computer seems the more so. Its capacity to summon from cyberspace in a fraction of a second a vast range of human knowledge (epitomised by that prodigious search engine Google) far outstrips the competence of the human mind. But for all that, the most powerful computer imaginable could never fall in love, or compose a line of poetry (knowing it to be poetry), or tell a joke, or smell a rose, or do any of the zillion banal things that fill our everyday lives. Nor will they. Thus the metaphor of 'brain as computer' must at some point cease to hold – but not before offering up many profound insights. We must first reflect, if briefly, on why that computer analogy is so compelling. The shift from the 'map' to the 'computer' metaphor entails a shift from investigating the brain's *macroscopic* structure, visible to the naked eye, to the *microscopic* – a shift anticipated back in 1872 when an Italian biologist, Camillo Golgi, 'discovered' the harmoniously beautiful and delicate tracery of the neuronal connections to the brain.

The most striking feature of the internal structure of the brain when sliced in half is the obvious distinction between the inner core of white matter and a wafer-thin, convoluted layer of grey matter on the outer surface. The brain is full of surprises, but few are more striking than the fact that the white matter consists entirely of nerve fibres connecting one part of the brain to the other – thus everything the brain *does*, all its seeing, thinking, emoting, rationalising, the entire sum of our mental life, 'happens' in that wafer-thin crinkled veneer of grey matter an eighth of an inch deep which, uncrinkled and ironed out, would cover an area no bigger than four medium-sized napkins.

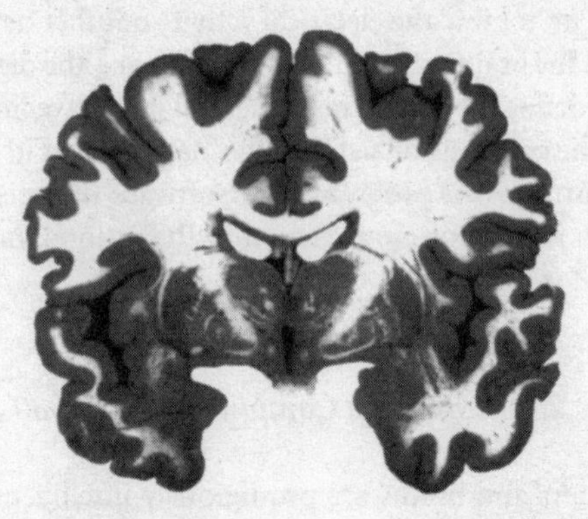

Fig 8-2: *This cross-section of the brain reveals the central core of 'white' matter, consisting of billions of nerve fibres variously connecting one part to the other, receiving sensory information and conveying instructions to the muscles. The sum total of mental life, all that the brain 'does', takes place in the thin layer of 'grey matter' on the surface, just one eighth of an inch thick.*

In the mid-nineteenth century, biologists peering down their microscopes had discovered the previously hidden landscape of the intricate structure of the tissues of the heart, lungs, bones, teeth, muscles – all revealed in the most intricate detail – but not of the brain, which appeared only as a pale, homogeneous mass. Then one day Camillo Golgi, who, caught up in the excitement of investigating this previously hidden landscape, had converted his kitchen into a small laboratory, inadvertently knocked a piece of brain tissue into a dish containing a solution of silver nitrate. When, a few days later, he retrieved the tissue and out of interest sliced it up and scrutinised it under the microscope, he was astonished to find that that pale, homogeneous mass had become, as if by magic, an extraordinary delicate tracery of neuronal connections. For some quite unknown reason, the silver nitrate stain had 'picked out' just one neuron out of a hundred from that impenetrable thicket, whose branching interconnections became visible with the same clarity as a tree on the skyline on a winter's day.

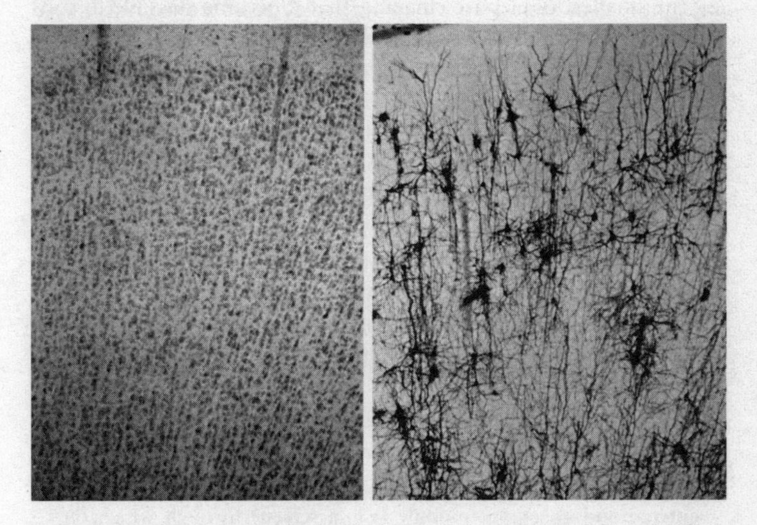

Fig 8-3: *The staggering complexity of interconnections of the neurons of the brain is illustrated by these two successive microscopic sections through the grey matter of the visual cortex of a rabbit. The section on the left stains up the nucleus (the black dot) of thousands of individual nerve cells, that on the right (stained by the Golgi method) just one in a hundred – each with their numerous connections to each other. The similarity between this rich, dense circuitry of the brain and the layout of a microprocessor 'chip' suggested its workings might be analogous to that of a computer.*

'One look was enough. Dumbfounded, I could not take my eye from the microscope. All was as sharp as a sketch with Chinese ink on transparent paper,' commented the Spanish microscopist Ramon Cajal a few years later. 'And to think this was the same tissue on which one could stare for ever, fruitlessly, baffled to unravel the confusion. Here on the contrary all was as clear and plain as a diagram.'

Subsequently, biologists with microscopes of ever greater magnifying power would discover that those individual neurons connect up with thousands of others across the narrowest of gaps or *synapses*, generating an immeasurably vast number of connections. Camillo Golgi could scarcely have appreciated the full significance of his discovery, as the miniaturisation of electrical circuits – which that delicate tracery of neuronal connections so closely resembles – lay far in the future. It was not until the invention of the transistor in 1948, capable of boosting the smallest of electric currents, that it became possible to conceive how the circuitry of the brain – with its billions of neurons (approximating to the number of trees in the Amazon rainforest) and trillions of synapses – might work like a computer.

Soon after, in the mid-1950s, Herbert Simon and Allen Newall, working for the Rand Corporation in America, devised a computer programme that could prove logical theorems in a way that resembled human performance, at which point it became possible to conceive of that vast interconnecting network of neurons in the grey matter as the ultimate computerised information-processing machine.

So one arrives at the 'computational theory of the mind', as lucidly summarised by philosopher and neuroscientist Raymond Tallis:

> A computer is a machine for processing what is, in a very extended sense of the term, called 'information'. In order to do this it has to have an *input* module (e.g. a keyboard) where the information can be entered and an *output* module (e.g. a screen) through which the outcome of the computer's operations can be expressed. Between the two is the central processing unit (CPU) where the processing takes place...[which] operates on material currently entered into it and also upon stored material previously entered into the computer and held in 'memory'.
>
> The analogy with the human brain is very persuasive, indeed inescapable. The input modules are the senses, eyes, ears, temperature receptors etc., the outputs are the visible movements and invisible physiological and biochemical adjustments that take place in adaptive response to those inputs. The highest levels of the brain constitute the central processing unit and the memory modules in which previous inputs and programmes are stored. The manner in which the neurons are connected corresponds to the wiring on circuit boards and so on.

Nowadays we are so accustomed to the notion of the brain as some form of glorified computer that it can be difficult to imagine the exhilaration of those early pioneers who first saw the possibility of creating these intelligent machines which would also, for good measure, explain the workings of the mind. The brain's capacity for processing information is phenomenal. 'We don't just have the power of a single computer in our heads,' writes communications expert Charles Jonscher. 'The true comparison would be a figure more like twenty billion computers. The complexities involved are genuinely difficult to imagine.' The brain, it is estimated, has the potential to form a quadrillion computations per second, which far outstrips the capacity of even the most sophisticated super computer to 'compute' its action. But the value of the computer metaphor, as with 'the map', lay less in it being a direct analogy than in allowing for the possibility that the profound mystery of the relationship between brain and mind might, at least in part, be soluble – where the two major determinants of who we are, nature and nurture, reflect respectively (if not precisely) the 'hardwiring' of the attributes of the human mind into the neuronal circuits of the brain at birth, and their elaboration by the programming 'software' of experience and upbringing.

Nature – 'Hardwiring' the Brain

The notion of human individuality as a judicious mixture of 'nature' and 'nurture' stretches back to the late nineteenth century and that already-encountered Victorian polymath (and cousin of Charles Darwin) Francis Galton. 'The phrase "nature and nurture" is a convenient jingle of words,' he wrote, 'for it separates out under two distinct headings, the innumerable elements of which personality is composed.'

Galton proposed the ingenious (if simple) method of 'twin studies' for assessing the respective contributions of nature and nurture, by comparing the character traits of identical twins (who share the same genes inherited from their parents) with non-identical twins (who do not). His findings were so definitive as to be almost embarrassing:

> 'There is no escape that nature prevails enormously over nurture,' he wrote. '[The identical twins] remain similar throughout their lives, not only in appearance but in ailments, personality and interests. [The non-identical twins], by contrast, grew more different as they grew older . . .

> my fear is that my evidence may seem to prove too much, and be discredited on that account.'

A century later, in the 1980s, the American psychologist Thomas Bouchard would refine Galton's technique, studying thirty-nine sets of identical twins separated at birth and brought up in very different social circumstances. They included Barbara, the adopted daughter of a gardener, and Daphne, whose adoptive father was a metallurgist. Barbara's first inkling of her twinship came when, checking her birth certificate to establish her entitlement for a pension fund, she noticed that the doctor had jotted down the time of her birth, which in Britain is used as a way of distinguishing between twins. She finally met her identical twin, after being separated for nearly four decades, at King's Cross Station in London.

> Each was wearing a beige dress and brown velvet jacket. They greeted each other by holding up their identical crooked little finger – a small defect that had kept each of them from ever learning to type or play the piano. They discovered they were both frugal, liked the same books, had been Girl Guides, chose blue as their favourite colour . . . both had the eccentric habit of pushing up their noses, which they called 'squidging'. They liked their coffee black and cold. They were both sixteen when they met the men they were going to marry. And both laughed more than anyone else they knew.

These astonishing similarities between identical twins reared apart attracted much attention. Bouchard sought to distinguish them from mere coincidence by conducting psychological tests that would reveal those features of character that could only have been genetically inherited. His findings were far in excess of anything Galton had anticipated, with over 40 per cent of the variation in character traits between individuals being due to genetic factors (the same degree of heritability as body weight), 10 per cent being due to the influence of family upbringing, and 25 per cent to fortuitous life events such as prolonged childhood illness, quality of schooling and so on. Bouchard had presumed that some traits would prove to be more heritable than others, but with few exceptions his findings proved almost boringly predictable, with virtually everything turning out to be heritable, and identical twins being much more similar than non-identical.

Galton was right. It would seem that a person's character is rooted in his or her genetic inheritance, 'hardwiring' the attributes of the mind into the neuronal circuits of the brain. 'We used to think our fate was in the stars,' James Watson, the co-discoverer of the Double Helix, remarked in a much-quoted line. 'Now we know in large part it is in our genes.' And it is not just character traits that are so hardwired: many uniquely human attributes – most obviously the faculty of language – were similarly inherited as 'hardwired' modules in the brain so as to permit children to pick up the syntax and grammar of language, where every word potentially has a multitude of meanings.

> 'For a child to learn the meaning of words without [a structure] is akin to an alien trying to discover the laws of nature by examining the facts listed in the census report,' observes the psychologist Lawrence Hirschfeld. 'Both would be doomed to positing thousands upon thousands of meaningless hypotheses...[which] would yield meaningful knowledge only rarely (if at all) and even then only by chance.'

Hence, the linguist Noam Chomsky had insisted (as already described) that children must be born with that 'language acquisition device', a module of neuronal circuitry by which they could make sense of the babble of sounds around them. The same consideration applies to music and mathematics and the ability to recognise faces and to understand that causes have effects, all of which would require their own module of specialised neuronal circuitry. And more still, for there is simply not enough information entering the brain through the eyes to enable it to define every aspect of the external world 'out there', its colours and shapes, or how objects are located in three-dimensional space. So again the brain must come 'hardwired' for vision – it must, in short, be born with an innate knowledge of how the world 'works'.

Silicon Valley's finest computer designers could only fantasise about the scale and precision with which the human foetus creates the necessary hardwired circuitry of its future mind, forming an average twenty-five thousand new neurons *a minute* over the nine months of its residence in the womb – a hundred billion in all, each connected to a thousand others, a staggering trillion connections in all. The precision of that circuitry, where the individual neurons migrate through the dense Amazonian rainforest of the brain to their 'connecting' destination, has been likened, by psychiatrist Professor Jeffrey Schwartz of the UCLA School of Medicine, to 'a baby crawling from New York to Seattle and winding up in the precise neighbourhood, in the right street at the correct house he was destined to reach from the moment he left Manhattan'.

No one can tell how this happens. The 'instructions' that guide those billions of neurons and form those trillion synaptic connections that 'hardwire' those character traits and modules for language and perception (and so much else) into the brain are, as we have seen, contained within a mere handful of just several thousand genes. There it is. The 2 per cent difference that separates the genomes of humans from their primate cousins instructs not just for a 300 per cent bigger brain, but an immeasurably more powerful and talented mind.

Nurture – 'Programming' the Brain

Now we turn to the second element of Francis Galton's 'convenient jingle', the formative influence of nurture in determining human individuality; or how (to pursue the computer metaphor) 'culture' in its broadest sense programmes that hardwired, genetically determined circuitry of the brain to ensure that while a little Texan will grow into an adult Texan, he would in other places and other times have become somebody entirely else. It is not however 'culture' that does the programming so much as the young brain that voraciously and remorselessly programmes itself, wrapping itself around and integrating into itself all that it encounters. This 'neuroplasticity' of the young brain, with the power to alter its own structure in response to the demands placed upon it, is seen perhaps most vividly in those instances where children born into one society, yet brought up in another, acquire its language, habits and values, irrespective of their own genetic inheritance. The American anthropologist Ashley Montagu reports how the children of early settler communities in the United States kidnapped by the indigenous population 'became so completely Indianised, it was only by being informed of their origins from others that they learned of their real extraction'. Similarly, when in the post-war years 100,000 Korean children were adopted by families in the United States in the largest ever trans-cultural adoption programme, they became completely 'Americanised' within a single generation.

This formative influence of nurture is self-evident, but its biological basis – *how* the youthful brain programmes, for example, the relevant syntax and grammar of English or Korean into its hardwired 'language module' – remained quite obscure until 1963, when David Hubel and Torsten Wiesel of Harvard Medical School sewed together the eyelids of the right eye of a newborn kitten, and observed the results six weeks later.

> We wanted to have a rough idea of [its] visual capabilities so we opened the closed eye and fitted the normal eye with an opaque contact lens. No elaborate testing of vision was necessary. When we put the kitten on a table it groped its way to the edge and tumbled onto a cushion we had placed on the floor. That is something that no self-respecting kitten would ever do. We took it as *prime facie* evidence that it was, for all intents and purposes, blind in the deprived [previously sewn-over] eye.

They then sacrificed the hapless kitten, sliced up its brain and were 'amazed' at the pallor of the nerve cells of the visual cortex supplied by the sewn-over eye – whose neuronal connections, when scrutinised under the microscope, had 'died away'.

There is more than enough in the findings of this simple experiment – so predictable, yet so astonishing – for a fistful of scientific theses. It was predictable because the profound influence of nurture presupposes that 'experience' must, as here, directly influence the physical structure of the brain. But it took the tangible evidence of Hubel and Wiesel's experiment to reveal what a truly astonishing phenomenon it should be that, for example, our ability to see is sculpted by the electrical impulses generated by the sights and sounds of the external world impinging on our senses and 'programming' the relevant neuronal circuits of the visual cortex. How, one might reasonably wonder, do those other formative influences of 'culture' – maternal love, sibling rivalry, books in the house – similarly engrave themselves on the neuronal circuits of the brain?

There is, it turned out, prodigiously more to the phenomenon of neuroplasticity than was revealed by Hubel and Wiesel's blind kitten tumbling off the table. Nurture can, if necessary, programme the properties of the mind *onto any part of the brain it wishes.* This emerged most clearly when in the early 1980s paediatric neurosurgeons at Johns Hopkins University Medical Center sought to cure children with intractable epilepsy by means of a 'last ditch' operation of cutting away the affected half of the brain. They anticipated that this might cause paralysis of the limbs or loss of speech, but on the contrary: 'We are always amazed,' one of the pioneers of the operation commented. 'Here they are running, jumping, talking, doing well in school . . . able to lead a normal life. The worst they suffered from losing half a brain was some impairment of peripheral vision and fine motor skills on one side of the body.'

The same phenomenal neuroplasticity is increasingly observed in apparently normal adults in whom a brain scan reveals some massive defect in its structure that could only have originated in infancy. Thus the brain scan of a previously fit and healthy lorry driver aged fifty-five, admitted to hospital in a coma after crashing into a tree, showed a vast, fluid-filled cyst occupying the front two-thirds of the skull, while 'his frontal lobes and a large portion of the parietal and temporal lobes were missing'. This remarkable facility of the young brain to compensate so successfully for the absence of these major structures presupposes, as Jeffrey Schwartz puts it, that the brain must be capable of 'the wholesale reorganising of neural real estate where parts intended to serve one purpose are redeveloped for another'.

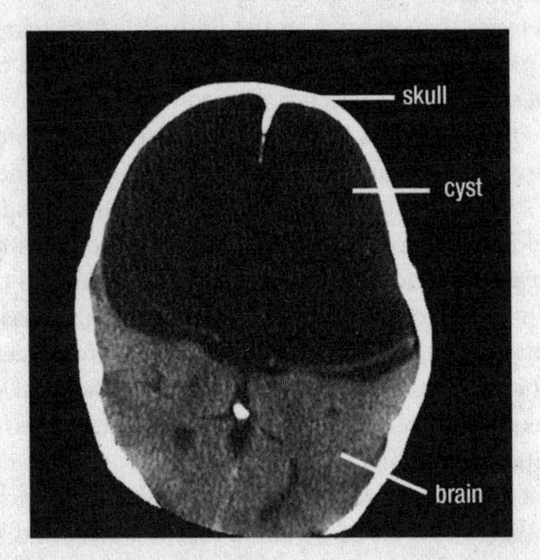

Fig 8-4: *The phenomenal power of plasticity. The brain scan of a fifty-five-year-old lorry driver born with a massive congenital cyst largely obliterating his frontal and parietal lobes. During infancy and childhood his brain would have reallocated their functions of reason, foresight and imagination to its residual 'real estate'.*

This neuroplasticity persists, in a more attenuated form, into adult life, where the brain responds to a catastrophe such as loss of sight by 'tuning up' the sensitivity of its other modes of perception. Thus, within a few weeks of the onset of blindness the part of the sensory cortex involved in touch has made significant inroads into the now redundant visual cortex, so that the fingers gliding over a page of Braille can 'read' the embossed dots on its surface. There is clearly more to this than the brain merely (!) accommodating to damage or injury. Like the commander of a city under siege seeking to maintain the integrity of its defences, the brain ensures its integrity as a functioning and coherent whole, responding to the surgically induced loss of the left hemisphere or the failure of the development of the frontal lobes by relocating the relevant properties of mind to the neuronal circuits that remain intact.

At this point the computer metaphor 'crashes' spectacularly. To be sure, the similarities remain immensely compelling: brains are prodigious information-processing machines, fashioned from comparable circuitry, while the neuron's two options of 'excitation' or 'inhibition' reflect the binary code of the manmade computer. And the more powerful and sophisticated computers become, the more human and intelligent they seem to be – epitomised famously when in 1996 the IBM computer 'Deep Blue' challenged the reigning world chess champion Gary Kasparov to a six-game match – and won.

But the computer's power to crunch numbers and work out chess moves is one thing, the capacity of the human mind to hold a conversation, or feel happy or sad, or indeed engage in any form of behaviour, is another. And here, when put to the test, the significance of that computer metaphor turns out to be precisely the reverse of that anticipated. The more closely the comparison is pursued, the more astonishing and uncomputer-like the brain appears to be.

The success of the digital computer lies in the rigorous logic of its operations, the ability to squeeze out any uncertainty, thus ensuring that the same steps will always lead to the same conclusion. The computer knows only a world of blacks and whites. It relies on its circuits being completely insulated from any source of outside interference that might affect the performance of its computations. The comparable process of the human brain, by contrast, is endlessly fluid as, moment by moment, it absorbs the avalanche of 'sensory input' pouring through its senses and adjusts its activity accordingly, scrawling its 'computations' across its living surface, which responds to the very act of writing.

This loss of confidence in the computer metaphor draws attention to how it has concealed, if unintentionally, the continuing profound ignorance about the most basic facts and the most elementary constructs of the working of the brain. First, it conceals the monumental differences in complexity between the circuitry of a computer chip with its dozen or so connections, when set against the thousands of synaptic connections of just a single neuron. That monumental difference in complexity is reflected too in their respective functions – for the tasks carried out by the most sophisticated computer, like Deep Blue, fade into insignificance when compared to the 'exploding explosion' of information processing involved in the brain being open and responsive to the amazing world around it.

> 'We are able to identify objects without prior warning or preparation (for example the face of someone not seen for many years) and make instantaneous absolute identification even when they are presented to us from angles and distances at which, and in lightings and settings in which, we have never encountered them before,' writes philosopher Raymond Tallis. 'This is totally beyond the reach of any conceivable computer, in which the relevant data area has to be mobilised and accessed before the process of "recognition" can begin.'

Next, the central enigma remains unresolved of how that monotonous electrical activity of those firing synapses translates into the limitless riches of the human mind. There has to be, it must be presumed, some 'code' hidden in its patterns of electrical activity to ensure that the firing of the visual cortex would give rise to those images of trees and birds as seen through my window, and the auditory cortex to those ever changing sounds of the birds' songs. There must, in addition, be another 'deeper' code that can separate out and distinguish between (for example) the colours of the rainbow, or the words, notes, pitch and volume of a Beatles song – and so on *ad infinitum*. But it is impossible to imagine what form those codes might take.

And the computer metaphor is deceptive also in the rather different sense of concealing the brain's proprietorial capacity to reassign, if necessary, its properties from one part of its real estate to another, to ensure its continued functioning as an integrated organ of diverse attributes.

The limitations of the computer model left the brain badly in need of a new metaphor. But what should it be? Karl Friston, neurobiologist at London's Institute of Neurology and a central figure in the third and closing part of this historical narrative of the unravelling of the brain, suggested, rather prosaically, the ripples in a pond.

> Standing by a pond in London Zoo, Karl Friston described a new vision of the brain to Harvard University psychologist Stephen Kosslyn: traditional thinking holds that the brain is some kind of computer, crunching its way through billions of inputs each second to output a state of consciousness. But really, the brain acts more as if the arrival of those inputs provokes a widespread disturbance in some already existing state. The pond, Friston suggested, gives you a better way of thinking about it. The brain is like a surface, its circuits drawn tight in a certain state of tension. You toss in a pebble – that's your sensory input – and you immediately get ripples of activity. The patterns say something about the way the pebble hit the surface, but they are mixed with the lingering patterns of earlier pebbles of input. And then everything begins echoing off the sides of the pond.

Friston's metaphor is qualitatively different in its intent from 'the map' or 'the computer'. It does not seek so much to clarify the details of the workings of the brain as to promote a radically different way of *thinking* about its workings – no longer merely responding to the external world with a blaze of activity of its fixed neuronal circuits, but possessed of an astonishing, evanescent fluidity that reflects its power to capture, moment by moment, the world 'out there'. We turn now to that 'new way of thinking' and to the technical developments of the PET scanner that, as described in the opening chapter, inspired it.

3. The Neuroscientific Revolution: 1980–2000

It is always difficult to discern the very few studies that will change the way we view the world among the flood of research that fills the scientific journals every week. But there is no mistaking the significance of the paper describing the very first investigations observing the brain 'in action' published in the journal *Nature* in February 1988. The title might seem obscure: 'Positron Emission Tomography (PET) Studies of the Cortical Activity of Single-Word Processing'; the authors, psychologist Michael Posner and radiologist Marcus Raichle of the Washington University School of Medicine in Missouri, were at the time unknown; and the phenomenon they investigated with their PET scanner – the pattern of brain activity when repeating single words – banal. But Posner and Raichle had, by incorporating for the first time the inner *subjective* private domain of human thought and language into the *objective* measurable domain of science, broached the seemingly insuperable barrier that had dogged the scientific investigation of the brain since the time of Broca – of scrutinising the brain not from the outside in, but from the inside out.

Posner and Raichle did more than just observe the brain 'in action'. They transformed neuroscience into 'Big Science', where the clean and sophisticated machinery of modern scanners and computers supplanted the traditional crude methods of brain research, with its overwhelming reliance on animal experiments carried out in fortress-like laboratories to avoid the attention of animal rights activists. Hubel and Wiesel's kitten with its eyelids sewn together had got off lightly when compared to the tens of thousands of experimental animals – cats, dogs, mice, monkeys and other primates – which had suffered the most terrible indignities in the name of science, strapped in restraining chairs, their skulls sliced open and stuffed with electrodes.

> Astronomers were building huge telescopes on the tops of remote mountains to peer at the origins of the universe. Engineers were sending rockets to the moon,' comments science writer John McCrone. 'But [brain researchers] felt forced to skulk in thc shadows, hoping not to attract too much notice. They could hardly invite the television cameras in to a recording session where an anaesthetised cat dangled in a steel frame, its lungs punctured to prevent the gentle motion of breathing from disturbing the position of the electrodes in its head; or to film the bumbling attempts of a monkey to insert its hand through a narrow slit several months after having had some small but crucial part of its motor cortex cut away.

The hardware of the new scanning techniques might be very costly – not just the machines themselves, which run to millions of pounds, but the legions of physicists, technicians, computer programmers and so on necessary to run them. But the investigations themselves, epitomised by Posner and Raichle's study, were not just 'clean' when compared to those animal experiments, but straightforward – conceived, conducted and written up within the few months that previously it would have taken just to wangle authorisation for yet another monkey experiment from a university's ethical committee. And the new methods of investigation had the further immeasurable benefit of there being something definitive at the end, in the form of a beautiful, multi-coloured scan of the brain 'lighting up' when performing one task or other, which was readily understandable by fellow scientists and the public alike.

Still, neuroscience would never have prospered if it had not delivered a whole series of fascinating and unanticipated insights into the workings of the brain – epitomised most forcefully by Posner and Raichle's very first paper on observing the brain 'in action'. They had asked seventeen volunteers to first read, and then recite, single words presented on a screen at the rate of one per second – and then to *think*, by linking the word with another with which it was readily associated (as in 'chair' with 'sit', or 'cake' with 'eat'). The PET scanner, as expected, had shown the brain 'lighting up', though not just in the speech areas as originally described by Pierre Paul Broca and Karl Wernicke, but across large tracts of its real estate. The more complex task of associating the word 'chair' with 'sit' (though in fact not complex at all) proved to involve in addition massive tracts of the frontal lobes, and the right hemisphere of the brain as well. This first, staggeringly simple experiment confounded all previous assumptions about the workings of the brain by showing how those anatomical parts, such as the language centre, with their discrete functions, were all summated, like 'the ripples of a pebble on a pond', into a dynamic, integrated network of electrical activity. And when those same volunteers rehearsed their word-association tasks so they knew their sequence in advance, and so could automatically generate the appropriate response, that blaze of electrical activity was extinguished and the networks fell silent. Who could have imagined that it would take millions of neurons, billions of synapses, to learn the simplest task, which once learned became so efficient and automatic as to show up as a mere flicker of activity on the brain scan?

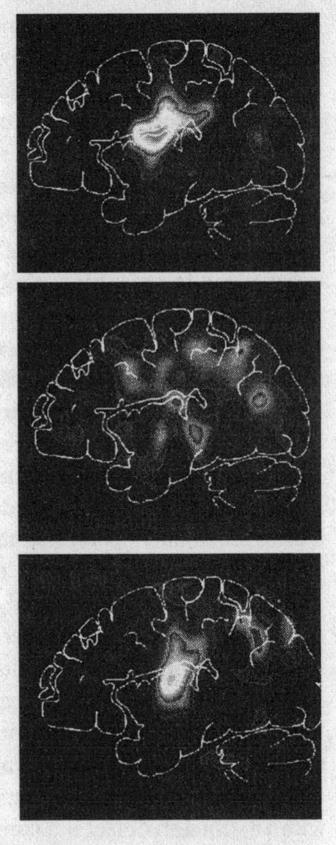

Fig 8-5: *The brain 'in action', learning a (very) simple task. The brain scans (top) of subjects asked to volunteer an appropriate verb (e.g. 'sit') for a list of nouns (e.g. 'chair') presented on a screen. The widespread and intense electrical activity generated by this task subsides with practice (middle), but resurges (bottom) when the subject is presented with a further list.*

As time passed, the methods for observing the brain in action became ever faster, more sophisticated, more revealing. Ten years on, in 1998, when Posner and Raichle organised a seminar under the auspices of the National Academy of Sciences celebrating the anniversary of their original paper, neuroscientists had deployed the scanning methods to investigate virtually every aspect of the human mind: perception, and vision in particular; language and its disorders; memory in its several forms, long- and short-term; pain and sensitivity; face recognition; conceptual knowledge and how the brain responds differently to expected and unexpected events – and much, much more. These investigations, Marcus Raichle observed, 'have the potential to provide unparalleled insights into some of the most important scientific, medical and social questions facing mankind'.

It would be a hopeless task to attempt to deal in any detail with the flood of scientific papers that poured out of the leading research centres around the world, but undoubtedly the most revealing, certainly in changing our understanding of the mind, fall under three main categories. The first concerns *perception*, and in particular how the brain, through the senses, captures the world 'out there', translating it into the electrical activity of those billions of neurons that create the forms, shapes, colours and sounds of our immediate experience.

Next comes *memory*, the linchpin of the human mind, with its ability to integrate the accumulated experience and knowledge of a lifetime into the neuronal circuits of the brain, to be readily recalled decades later. And finally the enigma of '*free will*', where non-material thoughts can nonetheless have physical effects, creating that human sense of agency by which we act on and influence the world around us. We will consider each briefly in turn.

The World Out There: Capturing an Instant

The most difficult of all scientific propositions to come to terms with is that the three-dimensional world we inhabit is not as it seems 'out there', but is created by the electrical activity inside our brains. Yet it must be so, for our brains, hidden away within the dark recesses of the skull, can have no contact with, no direct knowledge of, that objective world other than through the senses – and as we close off each in turn, close our eyes, block the ears, pinch the nose, so the colours, forms and sounds of the world drain away.

Further, and this is yet more difficult to credit, while we have the overwhelming impression that the greenness of the trees and the blueness of the sky are streaming through our eyes as through an open window, yet the particles of light impacting on the retina are colourless, just as the waves of sound impacting on the eardrum are silent, and scent molecules have no smell. They are all invisible, weightless, subatomic particles of matter travelling through space. It is the brain that impresses the colours, sounds and smells upon them. 'For the [light] rays, to speak properly, are not coloured,' wrote the great Isaac Newton in his *Treatise on Opticks*; 'in them there is nothing else than a certain Power and Disposition to *stir up* a Sensation of this or that Colour.'

And it is easy to demonstrate, as Newton pointed out, that the several million colours we experience are 'stirred up' within the brain, rather than streaming in from the outside, because when 'a man in the dark presses the corner of his eye with his finger, he will see a circle of colours like those of a peacock's tail'.

More than three centuries on, we are no more reconciled to this extraordinary proposition, and the daunting challenge for modern neuroscience is to explain how the brain constructs just an instant of our subjective experience of being in the world – or, to take a specific example, to explain the unity and intensity of feeling experienced by the famous writer and environmentalist Rachel Carson.

> One stormy night when my nephew Roger was about twenty months old I wrapped him in a blanket and carried him down to the beach in the rainy darkness. Out there, just at the edge of where-we-couldn't-see, big waves were thundering in, dimly seen white shapes that boomed and shouted and threw great handfuls of froth at us. Together we laughed with pure joy, he a baby meeting for the first time the wild tumult of the oceans, I with the salt of half a lifetime of sea love in me. But I think we felt the same spine-tingling response to the vast roaring ocean and the wild night all around us.

The senses are, by definition, 'sensitive', acute enough to hear a pin drop or to see the flickering light of a night fire at a distance of several miles; so acute indeed as to capture every nuanced detail of the physical world with its several million colour 'values', the even greater number of sounds of the musical repertoire, tens of thousands of different combinations of smell, and that incomparable subtlety of touch that can discriminate between a puff of wind or a single molecule of water upon the skin.

The seemingly insuperable problem posed by these very sensitive senses is to understand first how they translate the subatomic particles of sight and sound into the electrical activity of firing neurons, and then how the brain 'retranslates' those electrical impulses into that 'wild tumult of the oceans' of Rachel Carson's subjective experience. The conventional view, which is of course very persuasive, has always held that those 'dimly seen white shapes' of the waves are somehow captured by the eye and impressed, as on a photographic plate, onto the visual cortex at the back of the brain. But it is not so, and the first of the major insights of the modern scanning techniques is to show that in fact the brain 'creates' the image of that external world.

> 'As surely as the old system of beliefs was rooted in the concept of a labelled image of the visual world received and analysed by the cortex [brain],' writes Semir Zeki, Professor of Neurobiology at the University of London, 'the present one is rooted in the belief that an image of the visual world is actively constructed by the cortex.'

And why? The image-processing capacity of the eyes is far too limited to hope to reflect the prodigious subtlety and complexity of the world 'out there'. Those limitations are familiar: the eyes themselves are in constant motion; the two images, not one, that they generate are distorted, turned upside down; and so on. Thus the only way we can 'see' is by drawing on the brain's own vast store of knowledge, accumulated since infancy, of what the information conveyed by those electrical impulses of the brain is likely to *mean*.

> 'The visual stimuli available to the brain do not offer a stable code of information,' writes Zeki. 'The wavelengths of light reflected from surfaces change along with alterations and illumination, yet the brain is able to assign a constant colour to them. The retinal image produced by the hands of a gesticulating speaker is never the same from moment to moment, yet the brain must consistently categorise it as a hand. An object's image varies with distance, yet the brain can ascertain its true size. The brain's task then is to extract the constant, invariant features or objects from the perpetually changing flood of information it receives from them ... *the brain cannot therefore merely analyse the images presented to the retina; it must actively construct a visual world.*'

Perhaps the simplest way of grasping how the brain creates that visual image is to pursue the wavelengths of light generated by those 'dimly seen white shapes' of that evening by the seashore deep into the structure of Rachel Carson's brain. The pencil of light from the waves enters the eye as billions of photons of pure energy and zero mass, travelling at 186,000 miles per second. Their absolute number is maximised still further by the relaxation of the muscles of the shutter-like iris, and is then focused through the lens onto the receiving 'satellite dish' of the retina, whose hundred million cells each contain a hundred million molecules of the light-sensitive pigment rhodopsin.

Here begins the brain's 'quiet conversation with nature', for the individual photons with their almost immeasurably small amounts of energy are nonetheless sufficient to generate a succession of complex chemical reactions to create a minute electrical impulse. Collectively those impulses converge on the optic nerve, speed along its highways, traversing the entire length of the brain to find the visual cortex at the back. This marks the beginning of the divergence from Professor Zeki's 'old system of beliefs', which held that vision ended with the electrical charge generated by the photons of those 'dimly seen white shapes' being impressed on the photographic plate of the visual cortex. But the reverse is the case, as the PET scanners of modern neuroscience would reveal, for the visual cortex fragments that image like an exploding firework up and across the surface of the brain so its many components can each be analysed by numerous separate but interrelated maps of the visual cortex. The first port of call is the primary visual cortex, known as V1, that responds to specific 'lines' – vertical lines, horizontal lines, curves, corners, borders – capturing the shape and depth of objects in the field of vision. And then?

Professor Zeki speculated that two other distinctive features of the world 'out there', the colour and movement of those objects in the visual field, would similarly have their own visual areas – and so it turned out. He scanned the brains of volunteers as they examined a highly coloured Mondrian-type abstract painting, and identified a consistent 'hotspot' several millimetres in front of the primary visual cortex, in an area known as V4 (*see page 204*). He then repeated the experiment, looking for an area specialised for detecting movement by scanning the brains of subjects while they viewed a pattern of moving black-and-white squares, and identified another 'hotspot' designated as V5, and so on. 'We could hardly have hoped for more satisfying results,' he would later write. 'We had no notion when we started of how rich a source of information such studies could be.' And that was just the beginning. Further investigation has revealed the brain perceives the world 'out there' by fragmenting the visual image into thirty separate specialised 'maps' scattered throughout the visual cortex (*see opposite*).

There is much more to vision than just seeing: it also guides our every movement and action – and for that we need to go beyond perceiving the shape, colour and movement of objects to recognising what they are and how they are positioned in 'space'. Thus, for Rachel Carson the simplest of purposive movements on that stony seashore, such as picking a pebble from the shingle, requires her to estimate both its exact size and weight, and to calculate its distance from her outstretched arm.

The precise function of many of those thirty visual areas is not nearly as well described as are those for colour and movement, but it would appear that they separate out into two streams. Those that might clarify the question of *what* the pebble is (its shape, colour

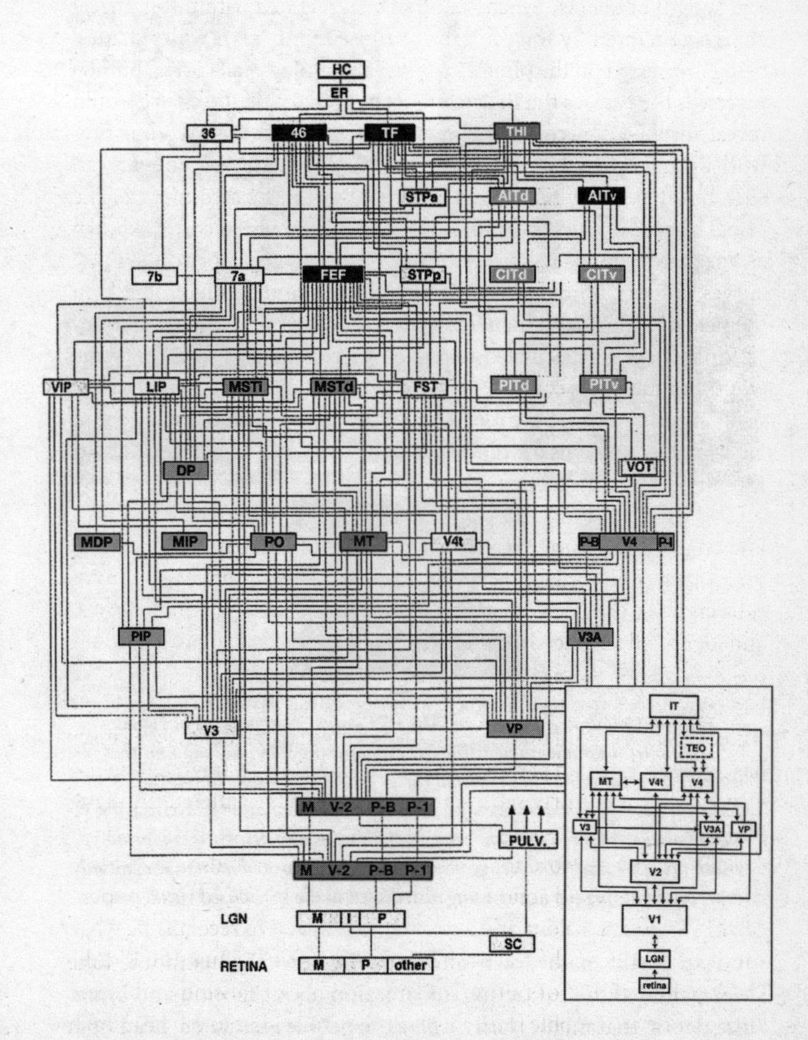

Fig 8-6: *The intricate circuitry of vision. This schematic portrayal (simplified in inset) of the network of connections of the visual cortex starts at the bottom, with the billions of photons impacting on the retina at the back of the eye. This generates the electrical firing of the optic nerve, that traverses the brain first to the primary visual cortex (V1), before being processed through a further thirty different sites, arranged in a hierarchy of fourteen different levels.*

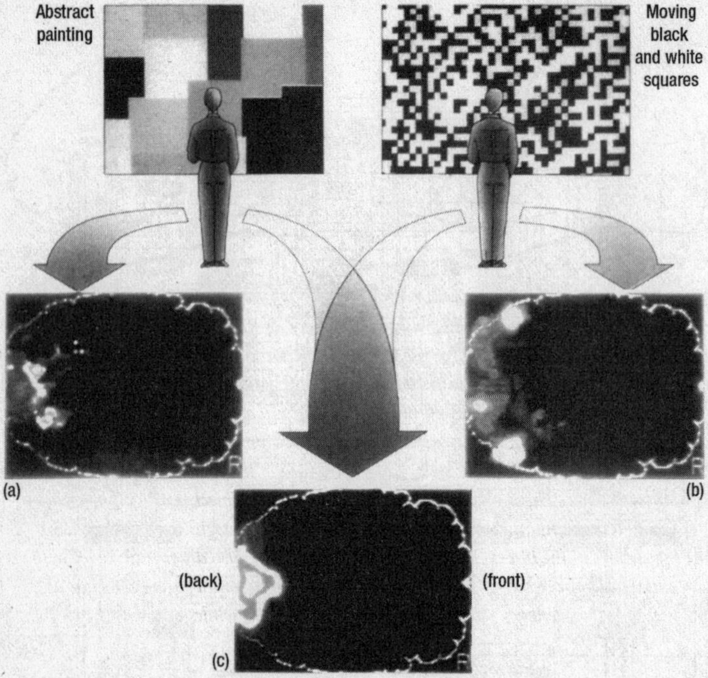

Abstract painting

Moving black and white squares

(a)

(b)

(back)

(front)

(c)

Fig 8-7: *Colour and movement. The PET scans of subjects when viewing a colourful Mondrian painting (a) pinpoint the V4 visual map for the interpretation of colour in the visual cortex at the back of the brain. By contrast the brain scans of subjects viewing a series of moving squares (b) activate the V5 visual area, concerned with movement. Both tasks also activated the 'primary' visual areas V1 and V2 (c), suggesting that both colour and motion are initially processed at this site before being distributed to the specialised visual areas.*

and so forth) pass down and laterally into the temporal lobe. Meanwhile, those extracting information about motion and binocular depth, that might clarify *where* the pebble is situated, head up in the opposite direction over the surface of the brain. The details are of no concern, but it is easy to see the markedly different patterns of activity involved first in recognising the pebble itself, and then locating it in space.

These findings do more than just delineate those 'what' and 'where' pathways which would allow Rachel Carson to pick up the pebble from

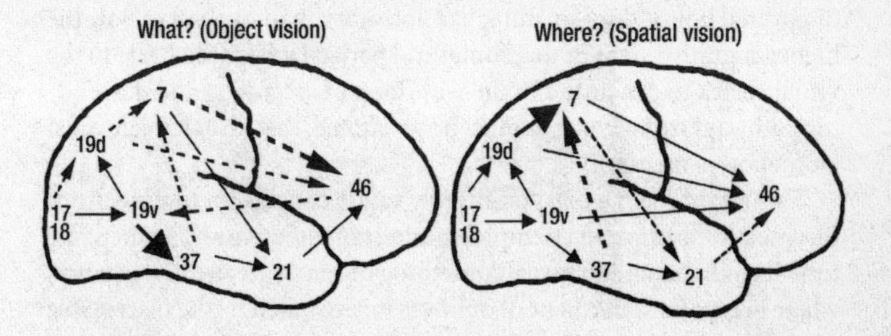

Fig 8-8: *What and where. The 'what' and 'where' of Rachel Carson's perception of the pebble on the shore are processed by two different pathways radiating from the visual cortex (areas 17, 18 and 19 on the left of the diagram) to the frontal lobes (area 46 on the right). (The numbers correspond to a conventional numeration for distinct parts of the brain.) The solid lines indicate a positive effect, the dotted lines a negative or inhibitory one, while the width of the lines reflects the strength of the influence. The main direction of flow for 'what' (object recognition) is down through the temporal lobe, while for 'where' (spatial recognition) it is up across the surface of the brain. The flow of electrical activity also differs markedly with the task. For 'what' the dotted line shows that the frontal cortex (46) has a negative feedback effect on the visual cortex (19), while for 'where' it is strongly positive.*

the shingle with such precision. They exemplify too the progressively mounting tension created by these new methods of investigation, which for all their novelty and interest (who could have imagined it would have worked out like this?) remorselessly compound the already substantial difficulties of explaining how we see. We have noted how at every stage the information received and interpreted by the retina and transmitted down the highways of the optic nerves to the visual cortex is less than sufficient to capture the world 'out there', only to find that that information has been further atomised into the extraordinarily complex network of visual maps and pathways. This leaves no alternative other than to suppose that the 'higher' mental properties of these integrative 'silent areas' of the frontal and parietal cortex must know how to reconstruct from those fragmented electrical impulses a coherent picture of the world 'out there'. This puts rather a different gloss on Professor Zeki's claims for the brain 'actively constructing' the visual image, which might imply that science in the recent past has discovered how it does so. But it has not, other than to observe how the higher cognitive areas of the frontal and parietal lobes 'feed back' to the visual cortex to generate a brain-wide loop of ascending and descending pathways from which, it must be presumed, that visual image must (somehow) emerge.

This might seem a disappointingly vague conclusion to draw from this most sophisticated attempt to understand how we see. At the same time it could be said to reveal something of much greater importance, where every single moment of our lives is permeated by the inscrutable mystery of our perception of 'being in the world' – and on two counts: the impossibility of reconciling that monotonous electrical firing of the neurons with the richness and detail of those booming and shouting dimly seen waves, and secondly the impossibility of knowing how the brain 'binds' it all together.

> 'This abiding tendency for attributes such as form, colour and movement to be handled by separate structures in the brain immediately raises the question of how all the information is finally assembled, say for perceiving a bouncing red ball,' observed David Hubel nearly thirty years after his original experiments sewing up the eyelids of kittens. 'It obviously must be assembled, but where and how, we have no idea.'

And that 'no idea' is '*no* idea'. We have no idea how Rachel Carson's brain, in less than 150th of a second, first deconstructed and then reconstructed those electrical patterns of activity generated by the minuscule forces of energy impacting on her senses – the colourless particles of light impacting on her retina, the soundless pressure waves impacting on her eardrum, the airborne molecules impacting on her olfactory nerve, the wind impacting on her skin. We have no idea how the brain translates those electrical patterns of activity into, respectively, the image of the silvery light reflecting off the ocean's surface, the crashing sounds of the waves, the smell of salt in the air, and the feel of the cold and wet of the night. We have no idea how they are integrated together into her clear, coherent, instantaneous sense of being in the world. We have, in short, no idea of the physical basis in the brain of every simple fleeting moment of her, or indeed our, life.

Remember This

Memory, the linchpin of the human mind, holds the past and present in a perpetual embrace – and also the future, for we draw on the experience of the past to plan for what is to come. It provides the immensely, richly detailed and ever-changing background to the drama of our lives. It allows us to learn from experience, and thus acquire the skills with which we negotiate the world in which we live. And it incorporates the influence of family and friends, education, upbringing and culture into our minds to mould our personalities and forge who we become.

This is as wide a brief as that for any component of the human mind, but nonetheless each of these attributes is underpinned by that truly remarkable property of the brain, as already touched on, its *plasticity*, its ability to wrap itself around the world and make it part of itself. The fundamental question memory poses for contemporary neuroscience is easily put – how do the neuronal circuits of the brain acquire knowledge (in its broadest sense) of the external world?

There is, as with vision, vastly more than one might suppose to memory, which comes in several different forms and whose three main components of *memorising*, *storage* and subsequent *retrieval* are intimately related to each other. Sticking with Rachel Carson on that stormy night, the 'life history' of a memory starts with the pencil of light from those 'dimly seen white shapes' traversing the brain to the visual cortex at the back. Here it is memorised, or encoded in the first staging post, '*short term*' or 'working' memory, where, drawing on information from elsewhere in the brain, it holds the image 'in the mind's eye' moment by moment. There it resides for an instant before vanishing like a puff of smoke, to be lost for ever unless there is some reason for it to be forwarded to one of several baskets of '*long term*' memory. At this point it seems the brain must distinguish (and it is a most extraordinary thing) between the many types of impressions held in the short-term memory-bank in order to determine to which of the several long-term destinations it should be forwarded. Thus it is customary to distinguish between *implicit* memories, those learned skills such as walking and talking that we just do without knowing why, and *explicit* memories that must be deliberately recalled. The brain must then distribute those explicit memories to two further baskets, the *episodic* or autobiographical, that would permit Rachel Carson several years on to recall that evening on the seashore with all its intensity, and the *declarative* fund of accumulated facts, her knowledge of, for example, the moon and the stars and the gravitational pull they exert on that vast tumult of the oceans.

This hierarchy of different forms of memory is both strictly compartmentalised and yet interdependent – as epitomised by the experience of Henry M, who following an experimental operation to cure his epilepsy that involved removing part of the temporal lobes, was no longer able to transfer his short-term into his long-term autobiographical memory, while his implicit memory (the ability to acquire new skills) and his recollections of the distant past remained intact.

> Henry works in a state rehabilitation centre, mounting cigarette lighters on cardboard frames, a task he has learnt to do skilfully,' writes the neurologist Colin Blakemore. 'But still he can give no account whatever of his place of work, how he gets there, or the type of job that he does. So Henry has not lost the ability to learn new skills of *movement*; he is quite simply unable to remember the new contents of his conscious experience. His general intelligence is not at all reduced and he is painfully aware of his own shortcomings. He apologises constantly for the absence of his mind. 'Right now, I am wondering,' he once said, 'have I done or said anything amiss? You see, at this moment, everything looks clear to me, but what happened just before? That's what worries me. It's like waking from a dream; I just don't remember . . . every day is alone in itself, whatever enjoyment I have had, and whatever sorrow I have had.

This strongly compartmentalised division of the several different forms of memory is not, it must be emphasised, some abstract or theoretical interpretation – but both real and essential. Indeed it is impossible, reflecting again on Rachel Carson's experience on that seashore, to imagine a memory system that would confuse her *autobiographical* recall of her impressions of the moon on that starry evening with her utterly different declarative knowledge of what it is and does.

There is more than enough evidence in the experiences of Henry M and others like him to suppose that these distinct forms of memory must be allocated to distinct parts of the brain's real estate. But the process of shifting those memories from one department to another was profoundly mysterious, and would remain so until the opportunity provided by Michael Posner and Marcus Raichle's scanning techniques of the late 1980s. Now for the first time it was possible to observe that cycle of encoding, storing and retrieving memories *as they happen*. But that was not all, for simultaneously Professor Eric Kandel of Columbia University's research into the memory circuits of the sea snail *Aplysia* was reaching fruition, seeking to account for that fundamental issue of 'what happens' when a memory is laid down in the circuits of the brain; how is a transient moment of the present permanently 'encoded' in the physical structure of the neurons, to be summoned at will from that vast storehouse of memories forty or more years later?

The PET scans of 'memory in action', together with Professor Kandel's *Aplysia* studies, are clearly complementary, two very different forms of scientific investigation that might catch the faculty of memory in a pincer movement and compel it to reveal its secrets. We start with Posner and Raichle's very first brain-scanning investigation – already touched on – whose findings would set the precedent for all that was to follow.

Posner and Raichle had assumed, reasonably enough, that scanning the brains of subjects performing the simple task of word association (linking 'chair' with 'sit') would reveal 'hotspots' of activity in the language centre in the left hemisphere of the brain, the focus of that form of explicit memory that registers the semantic meaning of words. And indeed their brain scans revealed just that, together with, as will be recalled, astonishingly and unexpectedly, the activation of vast tracts of the two main 'silent' and integrative areas of the brain – the frontal lobes and the parietal cortex. This was really the first indication that the brain functions as an integrated unit, activating billions of neurons when performing the simplest of intellectual tasks – in this case holding the word 'chair' in its short-term memory and searching to retrieve an associative term. But it took just a few moments of rehearsal, as will also be recalled, learning the list of words and thus knowing the answers in advance, for the brain to fall silent. This too was a revelation. The contrast between the major effort involved in thinking or acquiring new facts, and the minimal effort involved in drawing on what was already known, was the first graphic demonstration in miniature of that phenomenal plasticity of the brain – its ability to incorporate the sights, sounds and influences of the external world into its neuronal circuits. The same process is evident in acquiring that 'implicit' memory of knowing 'how to do things' – where the cabbage-shaped coordinating centre, the cerebellum at the back of the brain, 'lights up' in a blaze while the task is being learned (say, a child learning to tie his shoelaces), and falls silent when the sequence of movements becomes automatic.

Thus, right from the start, Posner and Raichle's brain-imaging studies, for all their apparent simplicity, promised the same sort of radical shift in the scientific understanding of memory and learning as Semir Zeki had done for perception. The next and obvious form of investigation was of memories themselves: where were those vivid pictorial representations of childhood holidays stored, or those long-remembered hymns and snatches of poetry? The brain-imaging studies offered the opportunity for the first time of pinpointing the precise whereabouts of the storehouses of those specific memories. Once again, their findings would undermine all previous assumptions about the workings of the brain.

That concept of a storehouse evokes an image of memories neatly stacked away, each in its own drawer, available for instant recall – as would seem to be the case with two types of memory in particular: the recognition of faces, and 'spatial topography', the sense of knowing where things are. The uniqueness of the face, along with the intonation of voice, is one of the great markers of human individuality, and its prompt recognition is the basis of virtually all our social interactions – as the writer and physician Oliver Sacks describes in his famous account of a man whose neurological deficit caused him to 'mistake his wife for a hat'.

> Dr P was a musician of distinction, well known at the local school of music as a teacher. It was here, in relation to his students, that certain strange problems were first observed. Sometimes a student would present himself, and Dr P would not recognise him; or, specifically would not recognise his face. The moment the student spoke, he would be recognised by his voice. Such incidents multiplied, causing embarrassment, perplexity, fear – and sometimes comedy.

Sure enough, researchers scanning the brains of volunteers as they distinguished between photographs of the famous (Bill Clinton) and the less so, found a distinct part of the frontal cortex would 'light up' that was clearly specialised for facial recognition.

A London taxi driver, by contrast, employs a different type of visual memory of spatial topography, the 'knowledge' of the layout of the streets and squares of the capital. Scanning the brains of taxi drivers as they rehearsed the route they would take from Grosvenor Square in Mayfair to Elephant and Castle in South London revealed 'hotspots' in another specific part of the brain, in the region of the right hippocampus.

This was all very impressive, but it soon emerged that these forms of visual recognition, with their own dedicated neural circuits, were highly exceptional. Both those very distinct forms of explicit memory, the autobiographical and the declarative (knowledge of facts), do not just light up (by now predictably) large tracts of the cerebral cortex, but in a remarkably similar distribution – as emerged in a study where volunteers were asked to reflect on and answer a series of questions tailored to their past experience, as in:

- Autobiographical events: You were the Christmas star in the school nativity play
- Autobiographical facts: Winkey and Frawley were friends at school
- Declarative knowledge: Cox's orange pippin is a type of apple

The great merit of the brain-scanning methods is their simplicity and clarity, capturing in a single image some profound (if extraordinary) truth about the brain's workings. But nowhere more so than where memory is concerned. It requires no specialist or technical expertise to 'see' in those black-and-white splodges of activity (*see overleaf*)from front to back of the left hemisphere of the brain how these very different types of memory nonetheless involve similar pathways of the brain. These images cannot contradict the supposition that the autobiographical and the declarative are quite distinct forms of memory – as they so clearly are. Nor can they contradict the supposition that specific memories, such as appearing as the Christmas star or knowing that Cox's orange pippin is a type of apple, are not 'laid down' and encoded somehow or somewhere in the brain. But they do most unequivocally deny that common perception of the brain as a storehouse of distinct memories, each assigned to a specific neural circuit, and forcibly bring to our attention the inadequacy of such a concept – even if there is nothing to replace it with. Put another way, these distinctive types of memory do not each have their own 'home address'; on the contrary, the *working*, *short-term*, *autobiographical* and *conceptual* forms of memory share many of the same neuronal circuits.

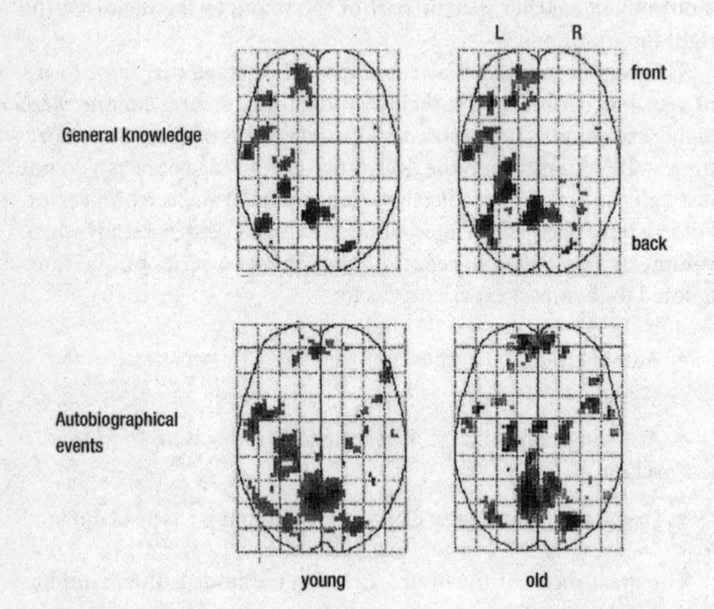

Fig 8-9: *Overlapping memories. A simple memory task lights up large tracts of the brain in both young and old, but there is a striking degree of overlap in the pathways involved in the retrieval of general knowledge (top) and the very different recall of autobiographical events (bottom), particularly in the frontal lateral and posterior parts of the brain in the midline.*

'It is worthwhile to consider why we are so dumbfounded,' writes Randy Buckner of Washington University in St Louis. '…Our many theories and ideas about what it is we are doing when information is retrieved from memory have provided little guidance on their more specific contributions.' There is every reason to be dumbfounded, as those seemingly fluid neuronal circuits contradict memory's most salient feature – its fixity over many years.

Subsequently, neuroscientists in Toronto would discover something more perplexing still – that in our lifetime our memories are reallocated from one part of the brain to another. Again, the task couldn't have been simpler. They invited a dozen 'young' and 'old' subjects, with an average age of twenty-six and seventy respectively, to first memorise and then recall word pairs (such as 'parents' and 'piano'). Both groups were equally proficient, but an analysis of the 'connectivity' of the network of pathways linking one localised area of

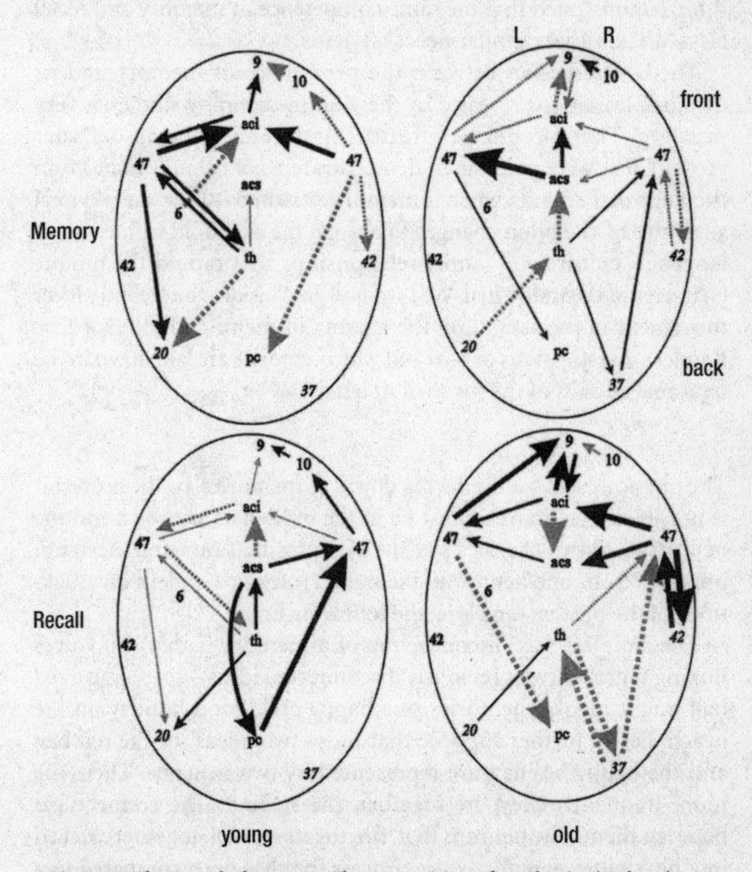

Fig 8-10: *The spread of memory. The* memorising *of word pairs (such as 'parents'–'piano') and their recall are scattered diffusely throughout the brain in quite different parts for young and old – illustrated here by the same schematic method as in the image on page 205.*

brain activity to another showed that they were not 'hardwired' at birth, but were constantly reassigned during a lifetime. Whereas in the young the predominant pattern of brain activity when memorising is in the left frontal lobe, and their recall is from the right (itself an astonishing observation), the same functions in the old are distributed equally between both sides. This is profoundly baffling, not least to Robert Cabeza, who devised this method of 'network analysis', who observed that these changes in the brain's connections over time demonstrated that the same competence in memory and recall tasks 'do not imply similar neural systems'.

The contradiction between the persistence of memory and its dynamic fluidity, as revealed by these brain-scanning studies, is very puzzling. There would be a reassuring feeling of being on 'safer ground' if it were possible to demonstrate *what actually happens* in the neuronal circuits when a memory is stored – how the physical structure of the brain changes to absorb the manifold influences of language, culture and family relationships that mould the unique character of the individual. We turn now to consider the second pincer movement in the assault on the enigma of memory, Professor Eric Kandel's investigations of how and why memories are laid down in the neuronal circuits of the sea snail *Aplysia*.

The physical basis for the 'laying down' of memories, or the nearest it is possible to get to one, must lie in the individual neurons, and the neurotransmitter chemicals at the synapses that transmit electrical impulses from one neuron to the next. Professor Kandel's investigations of the process ran along the following lines.

The simplest way of conceiving of a memory is that it involves linking together two previously disconnected ideas – so, seeing a red ball might invoke memories of a happy childhood holiday on the beach. Let us further suppose that those two 'ideas', of the red ball and the happy holiday, are represented by two neurons. Then, the more frequently they fire together, the stronger the connections between them. The neurons that 'fire together, stick together', locked in a physiological embrace – a process that has been compared to a lorry travelling up and down a dirt road, creating deep ruts in its surface.

From the 1970s onwards Professor Kandel explored this tentative (if, of course, simplified) hypothesis of the processes involved in memory and learning in the giant sea snail *Aplysia*, almost a foot long and known to the Romans of antiquity as *lapin marius*, because when sitting still it resembles a rabbit. *Aplysia*, despite its generous size for a snail, is blessed with a very small brain, compensated for by having very large (the largest of any living creature) neurons, twenty thousand in all. This combination of a simple brain and large nerves makes it an ideal experimental model for investigating the simpler forms of learning – such as the 'conditioned response' of the Russian neurophysiologist Ivan Pavlov, who in the early part of the twentieth century 'conditioned' his dogs to learn that an imminent meal was signalled by the ringing of a bell, so that with time the sound of the bell alone was sufficient to cause them to salivate.

Professor Kandel 'conditioned' his sea snails by spraying them with water, prompting them to return into their shell or mantle – while simultaneously applying an electric shock to the tail. Before long *Aplysia* would retreat into its shell at the slightest disturbance – which in the world of sea snails is the equivalent of learning something it did not know before. Kandel further identified in *Aplysia*'s very simple brain two sensory neurons where the modulatory sensory neuron conveying the sensation of the electric shock 'synapses with' that conveying the sensation of being sprayed with water, which in turn connect with the motor neuron that caused the muscles to retract its body into its shell.

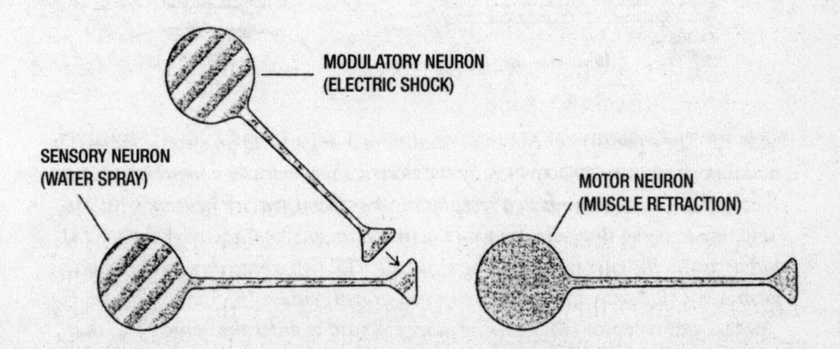

Fig 8-11: *Schematic outline of Professor Kandel's experimental investigation of memory in the sea snail* Aplysia, *which when splashed with water (sensory neuron) has been 'conditioned' by an electric shock (modulatory neuron) to withdraw into its shell (motor neuron).*

So what happens? The general drift is conveyed in diagram (a) below, whose perhaps discouraging chemical terminology can be safely ignored, as the significance of Kandel's findings lies in the process rather than the details.

This all seems very impressive, but this chemical cascade by itself cannot represent 'a memory', because the chemicals themselves are instantly broken down and promptly recycled. *Aplysia*'s learned response to that electric shock, by contrast, lasts its lifetime. This would require some *permanent* physical change in the synapse between the sensory and the motor neuron. So, along with everything else (*diagram b*), the protein kinase (5) must also penetrate

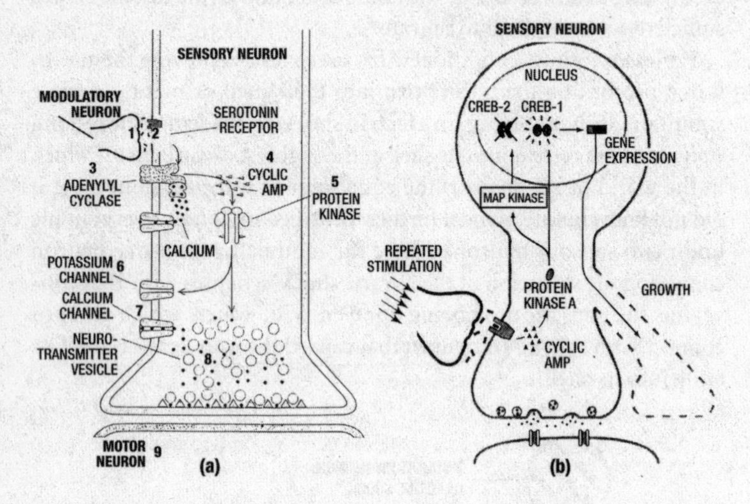

Fig 8-12: *The chemistry of* Aplysia's *conditioned response to an electric shock. The modulatory neuron (1) activated by the electric shock releases a neurotransmitter,* serotonin *(2), that binds to a receptor on the second sensory neuron, which in response to being sprayed with water activates an enzyme,* adenylyl cyclase *(3), that activates the energy-generating molecule ATP (4), converting it into its active form AMP, that activates another enzyme, protein* kinase *(5), that influences the* potassium receptor *(6), allowing more* calcium *to enter the neuron (7), that causes the release of the neurotransmitter* glutamate *(8), that crosses the synapse and lodges with the receptor in the motor neuron. This produces a further series of chemical reactions that feed back on the sensory neuron – but also cause the end purpose of all this activity: the muscles to contract, by which* Aplysia *retreats back into its shell.*

the nucleus of the nerve cell to activate the gene that codes for a protein that comes in two forms, CREB 1 and 2, which, in a way that is not properly understood, causes the sensory neuron to grow an extra synapse that strengthens its link to the motor neuron; thus, the more frequently the sensory and motor neurons fire together, the stronger the synaptic connection between them – and the simpler the stimulus necessary to generate a high level of activity. This process, known as long-term potentiation (or LTP), bears some resemblance to the findings of Posner and Raichle's original brain-scanning study, where rehearsal of that word association list ('chair' and 'sit') markedly reduced the levels of electrical activity in the brain by presumably strengthening the connections between the respective neurons (for 'chair' and 'sit' – if there are such), thus reducing the strength of the stimulus necessary to cause them to fire together.

'This broad biological unification' of the mechanism of learning and memory should, Professor Kandel hoped, 'accelerate the demystification of mental processes'. One might, more readily, come to precisely the contrary conclusion. The sea snail *Aplysia*, together with those PET memory studies, accelerates the 'mystification' by defining the limits of scientific understanding of both. There is, after all, nothing in *Aplysia*'s 'strengthened' synapse to tell us what its memory of that electric shock is, whether in the short term or long term, explicit or implicit, episodic or declarative. Or how humans might summon *their* memories in an instant from the deep recesses of the mind, where they may have been buried for forty years or more.

To be specific. To many, the flecks of black against a white background, as shown in the figure overleaf, might seem initially to be an abstract image, but as soon as it is pointed out, they are clearly recognisable as a dalmatian dog with the trunk of a tree in the background. Twenty years hence, anyone shown that same image again would retrieve it from the long-term visual memory bank, and recognise it instantly. Neither the findings of the PET scanner nor Professor Kandel's scientific explanations can begin to account for the power of memory to retain such visual images over decades and retrieve them at will, any more than they can account for remembering the words of a familiar hymn or recalling a telephone number – or Rachel Carson's memory of that 'wild night all around us'.

Fig 8-13: *The phenomenal power of visual memory. The mind is driven to impose the order of meaning on the perplexing patterns of light and shade presented to it, as here, in the black-and-white photograph of a dalmatian. Twenty years on, the interpretation of those scattered shapes of black and white, and their meaning, will be instantly recognisable.*

Free Will

The deep perplexities of perception and memory as revealed by the Decade of the Brain could scarcely be more fascinating and unexpected. But the most significant of all the attributes of the human mind illuminated (literally) by those imaging studies concerns 'the freedom to choose', otherwise known as 'the problem of mental causation'. Free will poses a 'higher order' difficulty (or challenge) when accounting for the many properties of the human mind in terms of the physical activity of the brain because it presupposes something that should in strictly scientific terms be impossible: where the *non-material* thoughts of the 'self' can so influence the function of the *material* brain as to compel it to pursue one course of action (like reaching out for a glass of wine) rather than another (not doing so).

The implications are as follows. Science holds that nothing can happen that is not governed by the natural laws of material causation. Thoughts are non-material, therefore by definition they can't cause anything to happen. Hence, my supposition that I am free to choose one course of action over another must be an illusion generated by the physical activity of the brain to create the impression that it is my non-material 'self', it is 'I', who is making the decision. Or as the historian of science William Provine puts it, 'Free will as it is traditionally conceived – the freedom to make uncoerced choices among alternative courses of action – *simply does not exist.*' And if that is the case, if free will is an illusion, then so is responsibility, because we cannot be held responsible for the physical activity of our brain. And if responsibility is abolished, then so is morality, because by definition a physical brain can have no knowledge of good or evil. And if that is the case, the whole notion of human action and activity would vanish – the saint and the criminal held in equal regard, for both would be no more than the stooges of their brain.

The only alternative proposition would be that the scientific doctrine is in error, and that it is indeed possible for my non-material thoughts to determine my actions. There is, as it happens, abundant scientific evidence that the non-material mind can indeed directly influence the material body: it is well recognised, for example, that asthmatics sneeze in the presence of plastic flowers, presuming them to be real, while meditators and others can voluntarily control 'involuntary' bodily functions such as heart rate and blood pressure.

The power of thought is seen when we change our minds. Some chance observation, a novel fact or insight previously unrecognised, and suddenly everything we know about a subject changes – as if a vast shudder has convulsed the contents of the mind, reorienting them in an entirely different direction.

But much the most direct everyday evidence for the power of the mind to directly influence the activity of the brain comes from the need to 'pay attention', that effort of will necessary to focus on one aspect of the world rather than another, on one thought in mind rather than another.

> 'No object can *catch* our attention except by the neural machinery [the brain],' observed the remorselessly profound late-nineteenth-century psychologist William James. 'But the *amount* of the attention which an object receives after it has caught our mental eye is another question. It often takes effort to keep the mind upon it. Though it introduce no new idea, it will deepen and prolong the stay in consciousness of innumerable ideas which else would fade more quickly away. *The whole drama of the voluntary life hinges on the amount of attention, slightly more, slightly less which rival ideas may receive.*'

A century later, the advent of brain imaging would allow James's masterful insights on 'attention' as the expression of 'the will' to be put to the test in two very different ways. There are few simpler brain-imaging tasks than to ask a group of volunteers to pay attention to one feature or other of an image, which, sure enough and predictably, reveals widespread activity throughout the brain. But it turns out that the mind 'focuses attention' by turning up or down the 'dimmer switch' on whatever is being attended to. Hold still for a moment and consider, first the greenery of the trees outside the window, then strain to hear the plaintive sounds of your neighbour playing the cello next door, and be diverted by the smell of the roast chicken drifting up from the kitchen. In parallel time, a brain scan would show markedly increased activity first in the colour, then the hearing, and finally the olfactory sectors of the brain, while simultaneously turning down the electrical activity of adjacent or competing parts of the brain. Thus the self 'pays attention' by 'sculpting brain activity', as neuroscientist Ian Robertson of Trinity College Dublin puts it, 'tuning up or down the rate at which the [relevant] sets of synapses fire'.

This might seem persuasive evidence that 'the mind really matters'; and it now appears that at a higher level still, just *thinking about one's thoughts* physically alters the neuronal circuits of the brain. The background is as follows. The two standard forms of psychiatric treatment are, first, drugs such as Prozac that alter the concentration and function of chemical neurotransmitters of the brain; and, second, psychotherapy, and in particular 'cognitive' therapy that seeks to change the way in which people think about and respond to their psychological problems. The practicalities of cognitive therapy in those with depression, for example, involve identifying ideas of low self-esteem and replacing them with a different, more positive set of beliefs.

This cognitive therapy might sound simplistic, but it has the great virtue of being highly, and consistently, effective. Its principles have proved readily adaptable to many types of psychological condition, including the well-recognised Obsessive Compulsive Disorder (OCD), where intrusive thoughts, frequently concerned with cleanliness and personal hygiene, take over the mind to the exclusion of everything else – or, as in a case described by Jeffrey Schwartz, where a young woman becomes obsessed about her partner's fidelity.

> Anna was twenty-four, a graduate student in philosophy . . . obsessed with the suspicion that [her partner] was unfaithful. Although she

never truly believed he was cheating on her, she was unable to stop obsessing about it. What had he eaten for lunch? Who were his girlfriends when he was a teenager? Did he ever look at pornographic magazines? Had he butter or margarine on his toast? The slightest discrepancy in his accounts set Anna off, making her whole world crumble under the suspicion that he had betrayed her.

This pattern of behaviour is strongly suggestive of some 'faulty wiring' in the circuitry of the brain. And sure enough, Professor Schwartz found that the brain scans of those with OCD exhibited increased activity in the frontal lobes and caudate nucleus in the mid-brain. He then repeated the scans after a course of cognitive therapy in which his OCD patients were encouraged to switch the focus of their obsessive thoughts onto some pleasing and familiar habit. Once again, no specialist expertise is necessary to interpret his findings of a marked reduction in the activity in those parts of the brain known to be involved in the perpetuation of those obsessive compulsive thoughts.

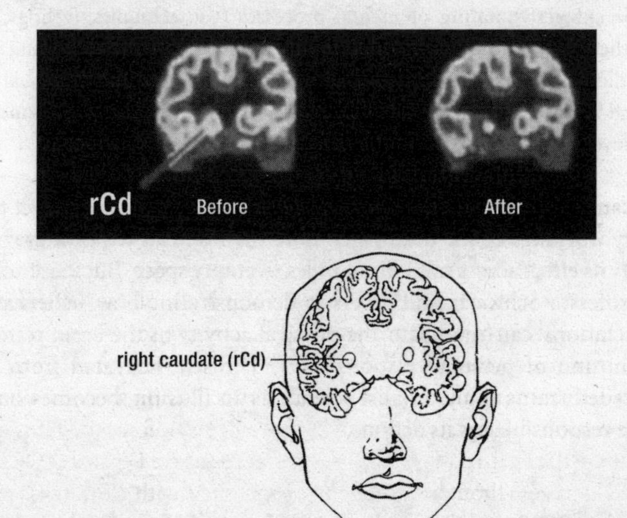

Fig 8-14: *Mind over matter. The potency of non-material thoughts to directly influence the physical workings of the brain is illustrated in these 'pre' and 'post' scans of the effects of ten weeks of cognitive therapy in markedly reducing the heightened activity of the caudate nucleus that is characteristic of patients with obsessive compulsive disorder.*

> 'This was the first study ever to show that cognitive therapy has the power to change faulty brain chemistry in a well identified brain circuit,' observes Professor Schwartz. 'We have demonstrated [the sort of] changes that psychiatrists might see with powerful mind-altering drugs – but in patients who had, not to put too fine a point on it, *changed the way they thought about their thoughts.*'

Professor Schwartz's discovery inspired numerous similar investigations, such as observing the 'before and after' PET brain scans in patients successfully treated for spider phobia (as shown by their willingness to stroke a tarantula), and other psychological disorders such as panic attacks and manic depression. The virtues of self-discipline proved another fruitful source for investigation by directly observing the power of the will to suppress feelings of sexual arousal when watching an erotic film.

> Collectively the findings of these studies strongly support the view that the subjective nature of mental processes (e.g. thoughts, feelings, beliefs) significantly influence the various levels of brain functioning. Beliefs and expectations can markedly modulate neurophysiological and neurochemical activity in brain regions involved in perception, movement, pain and various aspects of emotional process.

We cannot know *how* 'thinking about one's thoughts' can correct the faulty workings of the brain, any more than we can tell how gravity exerts its effect across millions of miles of empty space. But the studies of Professor Schwartz and others in demonstrating how 'beliefs and expectations' can 'modulate' the physical activity of the brain restore the notion of personal responsibility. The self, liberated from its degraded status in materialist science as an illusion, becomes once more responsible for its actions.

2000 and Onwards: the Rediscovery of the Soul

For the pioneer of PET scanning, Michael Posner, those astonishing images of the seeing, memorising and thinking brain 'in action' mark a defining moment in the history of science, comparable, he suggests, to Galileo turning the much simpler technology of his home-made telescope to the heavens and discovering the universe. His colleague Marcus Raichle concurs: addressing that tenth-anniversary symposium of their first paper, he argued that the neurosciences now offered 'unparalleled insights into the most important questions facing mankind' – which indeed they do, if not perhaps quite in the sense he intended. Rather, the most striking feature of the neurosciences, 'unparalleled' in any other field of scientific enquiry, is how each of the phases of the progressive unravelling of the secrets of the brain has been marked by a further deepening of the perplexity of its links with the spiritual mind.

The project of mapping the brain initiated by Pierre Paul Broca's discovery of the speech centre was predicated on the assumption that it would be possible to localise its many attributes – the senses, movement and language – to specific areas of the cerebral cortex. But for all that that has proved to be so, the dominant features of the brain remain its 'silent' areas, with their capacity to integrate and unify thoughts, sensations and emotions into a continuous stream of conscious awareness. How does the neuronal activity of those silent areas solve a problem or make a decision, or be uplifted by the emotions of joy and passion? Where and how do its neuronal circuits make reasoned judgements, or draw logical conclusions? How do they set the mind free to transcend time and space, and through the powers of creativity and imagination make sense of the human experience?

The next phase of the unravelling, prompted by the computer analogy, served to emphasise the brain's staggering powers of information processing while bringing to attention the twin unfathomable profundities of nature and nurture: first, how just a few thousand genes might instruct the arrangement of those billions of neurons with their 'hardwired' faculties of language and mathematics; and second, the physical basis of that all-encompassing property of neuroplasticity by which the brain incorporates into itself the experiences of a lifetime. Finally, the 'Big Science' of neuroscience observing the brain in action has revealed processes that defy all imagining: how every detailed nuance of the three-dimensional world is generated from within the dark recesses of our skulls, deconstructed and reconstructed within a fraction of a second; or how the brain categorises our memories into different 'baskets', shifts them from one to the other and somehow maintains them as a permanent record in those ever-changing neural circuits; or how, contrary to every known law of nature, non-material thoughts and emotions directly influence the physical structure of the brain.

Hence the paradox where the more we have learned from that great unravelling of the brain, the more elusive any general theory of its relation to the mind has become. Science may have a thorough understanding of the workings of the heart and lungs, but the unprepossessing three pounds of brain tissue confined within our skulls, like a vast intellectual black hole absorbs the most searching forms of scientific investigation. Its most elementary currency of a single thought defies any materialist explanation – let alone how that thought might be 'rational' rather than 'irrational', inspired rather than banal.

Once again the seemingly irresoluble conundrum of the relationship of the physical brain to the spiritual mind has resurfaced, escaping the confines of science to become, as philosopher John Searle describes it, '*the* most important problem: how do neurological processes of the brain cause those inner-first-person qualitative phenomena [of the mind]?'

The prevailing scientific view still holds. 'Conscious intelligence is a wholly natural phenomenon, the outcome of billions of years of evolution,' while those inner-first-person qualities are, claims Professor Paul Churchland of the University of California, 'nothing but' the 'interaction of nerve cells and the molecules associated with them'. The proponents of this supposed congruence of brain and mind draw heavily, if obscurely, on the computer metaphor. 'Conscious human minds are more-or-less serial virtual machines,' writes Daniel Dennett, 'implemented on the parallel hardware that evolution has provided for us.' We may suppose that the self is interpreting a never-ending stream of thoughts and emotions passing through its mind – but there is in reality nobody at home. The self, like free will, is an illusion generated by the physical activity of the brain to convey the impression that there is someone in charge. Dennett concedes that the details are unresolved of, for example, how the same monotonous firing of those neuronal circuits can give rise to such qualitatively different experiences as the smell of a rose or a Bach fugue, but sooner or later the onward march of science will decipher the code that translates one into the other, and then all will become clear.

Dennett's interpretation of the mind, argues Professor John Searle, annihilates the human experience – and unnecessarily so. The distinctive properties of brain and mind are, he insists, readily reconcilable by conceiving the mind as an 'emergent property' of the brain – just as the phenomenon of water in its various forms of liquid, ice and steam is an 'emergent property' of the arrangement of its molecules of hydrogen and oxygen atoms.

> '[The mind] is simply a higher-level or emergent property of the brain in the sense in which the solidity of ice is a higher-level emergent property of H_2O molecules when they are in a lattice structure,' he writes, 'and liquidity is similarly a higher-level emergent property of those same molecules when they are, roughly speaking, rolling around on each other.'

Searle's analogy is so obvious and compelling that one might wonder why it should not resolve, at a stroke, that brain–mind conundrum. But his interpretation has its own difficulties: for all the qualitative differences between water in its different forms of ice, liquid and steam, nonetheless the atoms of hydrogen and oxygen of which it is composed remain objective measurable physical entities. The brain and mind, by contrast, are different in essence – the brain is a physical 'thing', the mind and its thoughts are not. Further, Searle's 'emergent' interpretation falls foul of the numerous difficulties posed by the findings of the Decade of the Brain: how, for example, the monotonous firing of its neuronal circuits translates into that rich subjective world out there, or how those 'emergent' non-material thoughts can cause my hand to move so as to write one word rather than another.

The only certainty in all these attempts to grapple with the brain–mind conundrum would seem to be that 'we cannot know'; or, as the philosopher Colin McGinn puts it: 'The bond between mind and brain is an ultimate mystery, a mystery that human intelligence will never unravel.' This might seem a disappointingly inconclusive – if inevitable – finale to those prodigious attempts of neuroscience over the past century to provide a satisfactory scientific explanation of the mind in terms of the workings of the brain. It would, however, be wrong to suggest that we are simply left with the insurmountable barrier of Colin McGinn's 'ultimate mystery'. Standing back, it is possible to see how the (unanticipated) legacy of the Decade of the Brain has brought to our attention five cardinal mysteries of the mind that taken together offer the profoundest of insights into our understanding of ourselves – illustrated, for greater clarity, by reference to Rachel Carson's visit to the seashore on that stormy night.

(i) The Mystery of Subjective Awareness
The first mystery is how the fundamentally similar neuronal circuits in Rachel Carson's brain conjure from the barrage of colourless photons and soundless pressure waves impinging on her senses that vividly unique and unified sensation of that 'wild night all around us': the *sound* of the waves as they crashed on the pebbly shore; the *smell* of the sea spray; and the *feel* of the cold wind and rain on her cheeks.

(ii) The Mystery of Free Will
The second mystery is how Rachel Carson's intangible non-material thoughts fulfilled her desire to introduce her nephew to the wonder of the ocean by activating the synapses of her motor cortex, compelling them to fire in a logical and coordinated manner so as to first wrap him in a blanket and then carry him down to the beach 'in the rainy darkness'.

(iii) The Mystery of the Richness and Accessibility of Memory
The third mystery is how her brain captured the ecstatic experience of that evening and encoded all its detail as a permanent snapshot in the shifting, transient electrical activity of her neuronal circuits, to be summoned at will many years later. And not just the memory of that evening – but all the memories of her eventful life.

'The seemingly limitless and enduring capacity of human memory is a deep mystery in itself,' writes neurobiologist Robert Doty. 'It is this facility to sort with such alacrity and choice among the items of a lifetime, pursue in milliseconds obscure, half forgotten episodes and their cascading associations that presently defies credible clarification.'

(iv) The Mystery of Human Reason and Imagination
The fourth mystery is those 'higher' functions of reason and imagination of Rachel Carson's mind, through which she realised the significance of that evening – inserting her reflections, through her writings, into the minds of all who might read of it. She did, of course, much more, for in her book *Silent Spring* (1962) her *reasoned* thoughts, infused by a powerful moral sensibility, about the threat posed by chemical pesticides would

forge the environmental movement, thus placing her amongst the most influential people of the twentieth century.

(v) The Mystery of the Self

The fifth mystery is Rachel Carson's sense of self, that non-material being that seems so convincingly to be located just between and above the eyes, that both looks outwards to the external world yet presides over that inner life of subjective impressions and actions. That self, with its distinctive, passionate yet steely personality, would both change and mature as she grew older, yet remain the same – 'paying attention' all the time to the fulfilment of her life's work.

These may be 'mysteries' to science, but they are certainly not to ourselves. Indeed there is nothing we can be more certain of than the reality of our sense of self and our everyday perceptions of the world around us, our thoughts and memories. The paradoxical legacy of the Decade of the Brain, then, is to bring to our attention in the most forcible manner how the human mind, like the Double Helix, fails the test of scientific knowability not just once but twice over. First, science, for all it has revealed about the 'without' workings of the brain, can tell us *not an iota* about the 'within' of the non-material mind, nor how it imposes 'the order of understanding' by bridging that gap between those perceptions, thoughts and memories as we know them to be and the electrical activity of the neuronal circuits of the brain as they are known to science.

Next, those five cardinal mysteries as revealed by neuroscience correspond, it will be noted, with those five distinct integrated properties of the human soul as first described by Plato, elaborated on by early Christian theologians and refined by René Descartes. And that soul, freed of its theological connotations, is no mere construction of the human imagination, but rather a resilient entity that changes over time yet remains the same, and integrates into itself that single, coherent, never-ending stream of consciousness of the human experience of being in the world.

It is necessary, too, to appreciate why we know this to be so. The prestige of science is that its reliable methods of investigation can extend the range of human knowledge beyond the realms of the senses to comprehend the vastness of the cosmos and the intricate, microscopic complexities of the cell. But that scientific achievement fades into insignificance when set against the prodigious intellectual powers of the human mind to generate the most reliable knowledge of the world 'out there' in all its clarity and detail from the scantiest of clues generated by those colourless photons of light bombarding the retina and those soundless pressures of waves impacting on our eardrums. Thus, while we have good reason to infer from the recent scientific investigations of perception, memory and free will the reality of the non-material mind, that inference is a mere footnote when set against the *direct* knowledge we have of our spiritual inner selves. When the most certain thing I know is the reality of my non-material self as a unique, distinct, structured spiritual entity, then there is every reason to believe it to be so. And when I have the impression of myself as an autonomous being 'free to choose', then that is how it is, regardless of whether the ability of my non-material, freely chosen thoughts to influence my actions contradicts the laws of science.

The significance of neuroscience's inadvertent confirmation of the reality of the 'soul' becomes clearer still in the light of all that has gone before. We return first to the striking parallel with the findings of the genome projects, with the inference of that 'formative influence' through which the chemical genes of the Double Helix must impose the order of 'form' on the wondrous diversity of life. We then link together the rediscovered soul and that life force with Newton's law of gravity as the three forces of the non-material realm that impose order on the material universe and all within it. While, to be sure, we cannot know the nature of those forces, nor how they directly influence the material world of (respectively) neurons, genes and matter, we cannot dispute their power to do so without dismantling the whole structure of scientific knowledge of their *effects* – from Newton's recognition of the significance of that falling apple onwards. This in turn takes us back to that crucial moment in the mid-nineteenth century when science changed the direction of Western society by denying the dual nature of reality, of a material and non-material realm, and asserted instead the priority of its materialist view over the philosophic view of the world as we know it to be. But it is not so, for the astonishing cumulative legacy of the New Genetics and the Decade of the Brain is to have subverted the four fundamental pillars of that materialist view: that Darwin's evolutionary theory accounts for both the wondrous diversity of the living world and ourselves, and that the 'secrets' of life lie in the material genes, and those of the mind in the workings of the material brain.

Is this the case? Might further scientific investigation vindicate Darwin's mechanism of natural selection as the explanation for the infinite beauty and diversity of the natural world? Might biologists at some time in the future penetrate the impenetrable Double Helix to reveal how the genes of a snowdrop determine its delicate form and colour, so readily distinguishable from those of a tulip or any other form of life, or find in the 2 per cent of the genome that separate us from our primate cousins those random genetic mutations that gave rise to our upright stance and massively enlarged brain? Might neuroscientists discover how the electrical activity of the neuronal circuits of the brain gives rise to the non-material mind, and confirm as 'mere illusion' our perception of ourselves as free autonomous beings? The answer to all these questions must be 'no'. We are not just a mystery to ourselves, but our existence as the sole witness of the splendours of the universe is its central mystery. This enquiry closes by considering how science has so successfully asserted the contrary, by concealing from view those potent forces of the non-material realm and persuading us that we know so much more than we really do, or can.

9

The Silence

'Although many details remain to be worked out, it is already evident that all the objective phenomena of the history of life can be explained by purely materialistic factors . . . *Man is the result of a purposeless and natural process that did not have him in mind.*'

George Gaylord Simpson, The Meaning of Evolution (*1949*)

The world, and our understanding of it, is happily full of surprises – but few quite as surprising as the scientific revelations of the recent past. Who could have anticipated that the cell would turn out to be an 'automated factory' capable of 'carrying out almost as many unique functions as the manufacturing activities of man on earth'; or the profound implications of the necessity of the Double Helix to be simple, condensing the unfathomable biological complexities of living organisms within its two strands of chemical genes? Who could have predicted that the genetic instructions determining the staggering differences in form and attributes of fly and man should be fashioned from the same 'master genes'? And, most astonishingly of all, who could have supposed for an instant that the most significant insight of science's culminating projects of the New Genetics and the Decade of the Brain should be the (re)discovery of that 'dual nature of reality' that encompasses both the material world and the unfathomable power and profundity of that non-material animating life force and the human soul? These are 'no ordinary' discoveries. They introduce a further dimension into the common interpretation of how things are, where my thoughts and intuitions must be considered as every bit as 'real' as the chair on which I sit (indeed more so, for that chair, as

I know it to be through my senses, turns out to be in large part a construct of my mind). And yet one might be forgiven for supposing there is nothing in the least untoward or dramatic in these findings. It would be reasonable to expect that the findings of the genome projects would have prompted a critical re-examination of current theories of genetic inheritance. Not a bit of it. The journal *Science*, commenting on the recently completed genome of the sea urchin in November 2006, felt no need to draw attention to the remarkable finding that this creature, with its rather uneventful life (other than causing grief to unwary swimmers), has virtually the same number of genes as ourselves.

Similarly, it would be reasonable to expect scientists to have revised their exclusively materialist explanation for the human mind; but Britain's most prominent neuroscientist, Professor Colin Blakemore, insists:

> 'The human brain is a machine which *alone* accounts for all our actions, our most private thoughts, our beliefs. It creates . . . the sense of self. It makes the mind . . . we [may] feel ourselves to be in control of our actions, but that feeling is itself a product of our brain, whose machinery has been designed by means of natural selection.'

Professor Blakemore's allusion to the brain being 'designed by means of natural selection' brings us to the closely related and yet more surprising discovery of the recent past: the shocking inadequacy of biology's foundational evolutionary doctrine to account for anything other than the trivial differences that might distinguish one species of finch from another. That doctrine, we now see, is not merely flawed or incomplete, but its proposed mechanism of natural selection as the 'cause' of the diversity of living things is contradicted at every turn by the empirical evidence of science itself. Yet again, there is not the slightest hint that such insights might pose a challenge to Darwin's proposed evolutionary mechanism – which indeed has never been more loudly asserted as being 'as much in doubt as the earth's orbit around the sun'. 'Nothing in biology makes sense except in the light of evolution,' biologist Theodosius Dobzhansky remarked in 1973, and nothing, it would, seem has happened to challenge that interpretation.

> 'Darwin's theses of common descent, the gradualism of evolution and the theory of natural selection have been fully confirmed,' pronounced the doyen of evolutionary biologists Ernst Mayr in the journal *Scientific American* in July 2000. 'No educated person any longer questions the validity of the theory of evolution, which we now know to be a simple fact.'

Silence can at times be much more eloquent than words. Scientists are no more willing than anyone else to concede that their suppositions may have been mistaken, particularly so when (as here) the devastating challenge to their narrow materialist explanations comes from within the scientific enterprise itself. This might imply some deliberate conspiracy to conceal from public view the significance of the findings of the recent past. It would, however, be more accurate to suggest that scientists simply do not 'see' the wider implications – and why?

Science's unique contribution to human knowledge is its capacity to measure, quantify and analyse the material world of objects and physical forces, from the vastness of the cosmos to the minuscule cell. That commitment to materialism, as the biologist Richard Lewontin of Harvard University points out, is 'absolute':

> It is not that the methods and institutions of science somehow compel us to accept a materialist explanation of the world, but, on the contrary, that we are forced by adherence to materialist causes to create an apparatus of investigation that produces materialist explanations. Moreover, that materialism is absolute.

That absolute commitment to materialism has, as also noted, been immensely successful, culminating in that supreme intellectual achievement of weaving a single coherent account of the history of the universe, as described in the opening chapter. But along the way, that success almost inevitably fostered the belief among scientists that theirs is the *only* reliable avenue to getting at the truth of how things are.

There are, to be sure, many other forms of knowledge – common-sense knowledge, historical knowledge, knowledge of music and the arts, the tacit knowledge of skilled craftsmen or the clinical judgement of doctors, and much else besides. But scientists cannot acknowledge the possibility of there being a 'dual' nature of reality, with both a material and a non-material realm, for that would be to subvert their exclusive claims to understand how the world 'works'. Hence the silence. Scientists cannot 'see' the significance of the findings of the recent past because they cannot stand outside their materialist view and conceive of forms of understanding different from those in which they have been trained.

Science may have got along more than well enough (till recently) with its absolute commitment to materialism, but the same cannot be said of humanity, who uniquely inhabit the parallel, non-material domain. We live our lives engaging non-material minds like our own in conversation through the mutual exchange of non-material thoughts and ideas, entering into relationships with them, attributing the non-material values of good and evil to our experiences, invoking non-material concepts such as beauty and ugliness – none of which have any place in the language of science. We carry on doing these things as we always have, but our understanding of their significance is drastically narrowed. From Cromagnon man onwards the dual nature of reality was part of the commonsense of mankind – but no more. Nowadays, while children must learn in tedious detail the facts of the material world as revealed by science, such as the chemical composition of saturated fats, they gain not the slightest inkling of an insight into the mind-boggling attributes of the human mind. They must learn by rote the technical terminology of the early stages of foetal development, without the slightest intimation of the profundities of the forces that guide the process of embryonic transformation. The dual nature of reality has, in short, been censored, written out of the script as being of historical interest only, a relic of the superstitious ways of thinking of the distant past.

The eclipse of the non-material realm is an almost inevitable casualty of its close association with religious belief, where the wonders of the natural world and the unique spiritual nature of the human mind were commonly perceived as the most powerful evidence for a supreme intelligence. This disentangling of the precepts of science from religious belief might seem all to the good, for their confusion can only lead to mischief. It is much better, as the writer and biologist Stephen Jay Gould argues, that they should be seen as occupying 'non-overlapping domains', where 'science can no more answer the question of how we ought to live than religion can decree the age of the earth'.

Nothing at first sight could be more reasonable – except, of course, it is not like that. Science has extended its sphere of interest far beyond questions such as 'the age of the earth', and has asserted its own exclusively materialist interpretation of reality. Gould's reasonable view might be better summarised as: 'We scientists are very broad-minded people and have no objection to those who might find a purpose or meaning in their existence in one religious belief or another. That's fine, so long as all acknowledge science is the guardian of the flame of objective and rational knowledge, which just happens to include the scientifically proven certainty that man and everything else is the consequence of a known materialist process.' Or, as the evolutionary biologist George Gaylord Simpson puts it:

> Although many details remain to be worked out, it is already evident that all the objective phenomena of the history of life can be explained by purely materialistic factors ... *Man is the result of a purposeless and natural process that did not have him in mind.*

Gould's assertion of science and religion as occupying two non-overlapping domains could thus reasonably be interpreted as a deceptive sleight of hand to conceal science's 'absolute' commitment to materialism. The point at issue is not that science deals with 'facts' and religion with 'values', but precisely the reverse – the profound influence of science's materialist 'values' on Western thought over the past 150 years is grounded in the dismissal of (or failure to recognise) all those 'facts' that might point to the reality of a non-material realm.

Those materialist values are so pervasive, in the Western world at least, that it can be difficult to recognise their partiality, their incompleteness as an account of the human experience. They entail, in brief, the assumption that science has proven the world to be organised 'strictly in accordance with materialist principles'. Hence the supposition of the reality of the human soul must be a mirage, there can be no moral laws to guide our conduct, the notion of free will must be an illusion, and there can be no 'higher' meanings to our existence other than those that we invent for ourselves. Or as the philosopher Bertrand Russell puts it: 'There is darkness without and when I die there will be darkness within. There is no splendour, no vastness anywhere, only triviality for a moment and then nothing.'

It would seem necessary to reconsider briefly how and why those materialist 'values' should have eclipsed for so long the contrary 'philosophic' view, with its intimation of there being 'more than we can know'. This might seem a formidable task, but it is summarised in just a single concept, albeit the most important in framing the modern world: the concept of 'progress', whose significance is best appreciated by revisiting briefly, and from a slightly different perspective, the immensely influential ideas of the Enlightenment touched on in an earlier chapter.

We are so familiar with, and committed to, the notion of progress – technical progress, industrial progress, medical progress, social progress – that it can be difficult to recognise its higher, more 'metaphysical' connotations in giving meaning and purpose to human existence, the promise of a better world where all might enjoy the fruits of prosperity, freed of the burden of physical suffering and injustice. Progress is objective and measurable, its benefits readily apparent in every new technical innovation, captured in the statistics of social progress of the rise in literacy and the decline in infant mortality.

Progress, too, is virtually synonymous with the scientific enterprise, its arrival on the human stage being marked by the scientific revolution of Galileo and Newton who, inspired by the notion of human reason as 'the supreme authority in matters of knowledge', held out for future generations the promise that man might know anything he wanted. And it is that juxtaposition, between disinterested, objective, impartial scientific *reason*, and *faith*, the faith of Biblical revelation grounded in the countervailing authority of the Church, that more than anything else would define the future direction of Western thought.

> 'Science suddenly stood forth as mankind's liberation, empirical, rational, appealing to a concrete reality that every person could touch and weigh for himself,' observes Professor Richard Tarnas of the California Institute of Integral Studies. 'Verifiable facts and theories tested and discussed among equals replaced dogmatic revelations hierarchically imposed by the institutional Church.'

The findings of the scientific revolution would in turn inspire the Enlightenment, whose most famous son and undisputed leader, the French philosopher Voltaire, articulated the struggle for human freedom under the banner '*Écrassez l'infame*' (Crush iniquity) – the iniquity of a corrupt Catholic Church and the absolute power of the monarchs of Catholic Christendom. Forced into exile in Britain in 1726, Voltaire fell under the spell of the ideas of the scientific revolution presided over by the now ageing Sir Isaac Newton, seeing in his 'natural laws' of gravity and motion that govern the universe a model for human society. That society would be stable and harmonious, its citizens empowered by 'natural rights' – to a fair trial, freedom of expression, and protection from the tyranny of a capricious state. Newton's 'natural laws' suggested too a model for a new form of religious belief: 'natural religion', grounded not in the dogmatic tenets of Christian faith with its miracles and 'tales of demons thrown out' and all the rest, but *in reason*. Voltaire, despite his antipathy to established religion, held there to be nothing so self-evident as the two halves of human experience, of the material and the non-material, from which he could only infer the necessity for there being an 'eternal and supreme' God. The most powerful evidence for the divine hand was (as ever), first, the wonders of the natural world, from 'the meanest of insects . . . the disposition of a fly's wings or the feelers of a snail' – and second, the rational human mind.

> We are intelligent beings: intelligent beings cannot have been formed by a crude, blind, insensible being: there is certainly some difference between the ideas of Newton and the dung of a mule. Newton's intelligence, therefore, comes from another intelligence.

The values of the Enlightenment that Voltaire articulated so forcefully would be written into the constitution of the newly independent United States, and would pervade the struggles for democracy and universal suffrage of the nineteenth century. But Voltaire's 'natural religion' grounded in reason would founder in the face of the remorseless onward march of science, and of two technical innovations in particular that would drive deep inroads into that bastion of the non-material realm, the seemingly inscrutable mystery of self-renewing life and its wondrous realisation in the near-infinite diversity of the natural world. Those innovations were, first, the microscope, that would reveal the hitherto hidden and successively more fundamental levels of 'organisation' of living organisms; and secondly chemistry, whose analytic methods would reveal the chemical basis of life. Though it might seem late in the day, we must consider them briefly in turn, for their staggering success would define the direction of biology over the subsequent 150 years – which in the process would drive underground the reality of the non-material realm.

* * *

The origins of the microscope stretch back to the early part of the seventeenth century, when a Dutch spectacle-maker, Hans Lippershey, holding a spectacle lens in each hand and lining them up in the direction of the steeple of a nearby church, was amazed by the apparent proximity of the weathercock perched so far above the ground. He duly fitted the two lenses into a tube to maintain their relative distance from each other, and thus created the first telescope. This might have remained a mere curiosity were it not for the Florentine genius Galileo, who, hearing of Lippershey's discovery and realising its potential for resolving the astronomical controversies of his day, constructed his own telescope and discovered the millions of stars that make up the Milky Way, the rings of Saturn, the moons of Jupiter, and much else besides.

The same principles of magnification that permit the human eye to visualise the distant stars can equally be applied to enlarging the very small, and with a few technical modifications, Galileo's telescope became the microscope – though it would require considerable technical modification before biologists in the early part of the nineteenth century would begin systematically to investigate the micro-anatomical structure of every tissue and organ – the heart, lungs, kidneys, skin and liver, culminating in Camillo Golgi's astonishing discovery of the delicate tracery of the electrical wiring of the brain, as described in the previous chapter. This previously hidden world, with each intricate organ elegantly 'designed' for its relevant purpose, offered, as one of its pioneers, Jan Purkinje, remarked, 'an inexhaustibility of new possibilities' where '*almost every day* brings forth new discoveries'.

Along the way, in 1839, Theodore Schwann of Berlin University would formulate what would become biology's grand 'unifying' idea, that all living things are composed of the same fundamental unit – the cell – whose protean forms would subsequently be compared to 'a set of magic bricks':

> Some rigid as bone, some fluid as water, some clear as glass, some opaque as stone, some factories of furious chemistry, some as inert as death, some engines of mechanical pull, some scaffoldings of static support . . . each one of all the millions and millions so finely specialised into something helpful to the whole, as if each understood its special power.

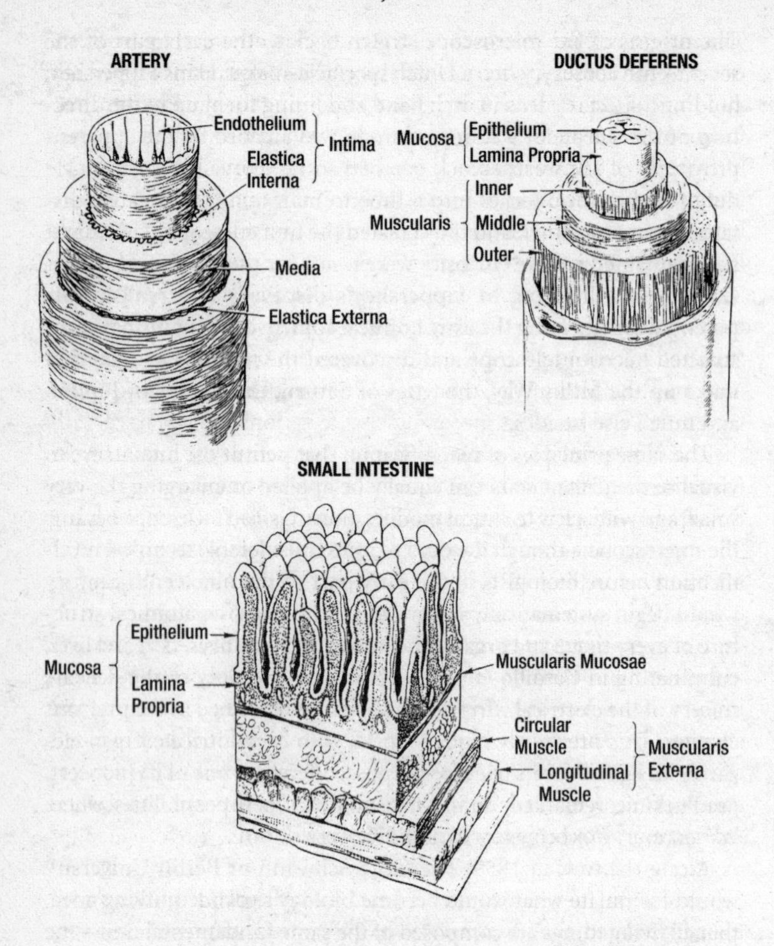

Fig 9-1: *'An inexhaustibility of new possibilities'. The nineteenth-century microscopists' investigations of the structure of tissues revealed numerous important insights. The diverse tubular forms of the gut, airways, ureters and blood vessels are constructed from relatively few elements, following the same basic pattern, but each is nonetheless uniquely suited for its purpose. The combination of the two layers of muscle fibres, the inner 'circular' and outer 'longitudinal', propels blood down the arteries, sperm down the ductus deferens and the gut contents through the small intestine. But their specific function is revealed by the form of the distinct inner layer, where the non-sticky walls of the arteries are ideally suited to the rapid transit of blood, while the fronds of the small intestine massively extend the surface area for the absorption of nutrients.*

The enthusiastic microscopists then turned their attention to the greatest and profoundest mystery of all: the astonishing fidelity with which all forms of life reproduce their kind, observing for the first time the fusion of sperm and egg at the moment of conception and the process of embryonic transformation into a fully formed organism. 'Nothing,' comments the historian of science William Coleman, 'could compare with [observing] the beauty of those rapid transformations working themselves out in front of their eyes.'

Paralleling this remorseless unravelling of life's hidden structures, the German chemist Justus von Liebig would resolve the further outstanding question of how the living organism transforms the inanimate matter of its nutrients into its own living flesh and blood – and the energy and heat that drives its vital functions. It all started with the chemical revolution of the late eighteenth century, initiated by the French genius Antoine Lavoisier, who, like Newton before him, penetrated far beneath the appearance of things to find the unifying scientific explanation of the seemingly inexplicable. 'Has anyone ever been able to explain how a log in the hearth changes into glowing embers?' Voltaire had asked in 1738. 'From the stars to the earth's centre, in the external world and within ourselves, every substance is unknown to us.' But no longer. Lavoisier, in a series of brilliant experiments, demonstrated that each of the supposedly fundamental elements of matter – air, fire, earth and water – was itself composed of the interaction of the same chemical elements: air composed of atoms of nitrogen, oxygen and carbon dioxide; fire and heat by the interaction of oxygen and carbon; water of atoms of hydrogen and oxygen – and so on.

Lavoisier achieved great things, but the scale of the facts waiting to be uncovered, the puzzles waiting to be solved, was simply too vast to be encompassed by any individual. They required a new type of scientific institution, the laboratory, first established in the universities of Germany at the close of the eighteenth century, where the vast possibilities of chemistry could be unravelled in an organised and systematic way. The first laboratory, and a prototype for all that would follow, was created by the young German Justus von Liebig, who in 1824, aged just twenty-one, was appointed Professor of Chemistry in the German city of Giessen – 'a formidable figure in a formidable scientific age'.

Fig 9-2: *A contemporary engraving of Justus von Liebig's first chemical laboratory in Giessen shows a hive of activity. The students' top hats were intended to protect their hair from the flying sparks given off by the charcoal fires necessary for their chemical experiments.*

The fascination of chemistry lies in its ability to find the common causal explanation linking the most seemingly disparate phenomena. And what could be more different than the two 'organic kingdoms', as they are known, of plants and animals? To be sure, they are both 'living', although one would be hard pushed to think that an apple had much in common with a tiger, or a rose bush with an elephant. But Liebig and his fellow chemists discovered to their astonishment when analysing tissues from the two 'kingdoms' that they were comprised of exactly the same 'stuff', variations on just three chemical compounds

– carbohydrates, fats and proteins – that were chemically strikingly similar, each with a backbone of carbon atoms running down the centre to which are attached atoms of hydrogen and oxygen, thus allowing them to be transformed one into the other.

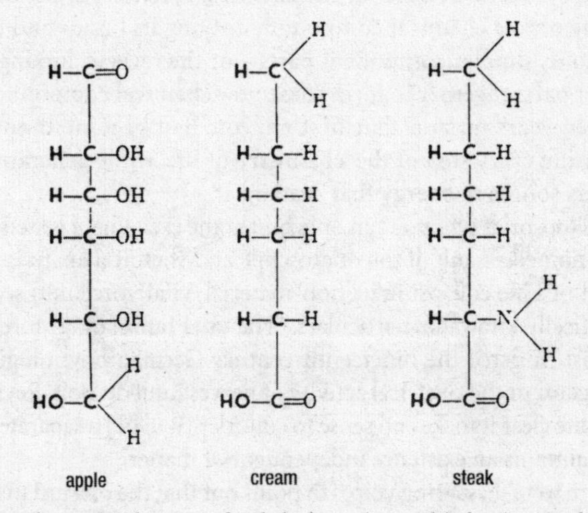

Fig 9-3: *The taste and texture of carbohydrates (an apple), fats (a pint of cream) and protein (a steak) could scarcely be more different from each other. But this schematic diagram illustrates their chemical structure, with a backbone of carbon atoms, is surprisingly similar – thus explaining how it is that the vast diversity of foods absorbed through the intestine and metabolised in the liver can be reconstructed to form the building blocks of tissues and organs.*

Suddenly and dramatically, the deep mysteries of the transformation of nutrients of food into flesh and blood were resolved. 'How beautifully and admirably simple with the aid of these discoveries appears the process of nutrition,' Liebig exclaimed in his major work *Animal Chemistry* (1842). But it was one thing to show the essential chemical similarity of the tissues of plants and animals, quite another to account for how they might first be disaggregated into their basic molecules in the gut and then reconstituted and transformed into the tissues and organs of the human body. Then in 1833 a couple of French chemists crushed germinated barley in cold water, added alcohol, and extracted a white precipitant that proved quite staggeringly potent – capable of breaking down two thousand times its own weight of starchy carbohydrate into component simple sugars of dextrine and glucose in just a few minutes. They had found the first enzyme, one of thousands of immensely complex proteins, each with a slightly different shape, designed to facilitate just one or other of the chemical reactions in the body by latching onto a chemical compound, untying its bonds and reducing it to its simpler component parts – or the reverse, linking those simpler parts together to form some new chemical compound. One hundred years on and that first enzyme had become thousands, facilitating every step of the 'chemistry of life' while generating the limitless source of energy that sustains it.

This too-brief synopsis can only hint at the excitement generated by the combined assault of the microscope and chemical analysis on the Citadel of Life, collapsing its non-material 'vital' force into so many scientifically knowable particulars. 'The vital [animating] force does not exist,' insisted the nineteenth-century German physiologist and investigator of the electrical activity of nerves Émil du Bois-Reymond. 'It is quite clear it makes no sense to consider [it to be] a separate entity that maintains an existence independent of matter.'

There were dissenting voices to point out that the onward march of biology's investigations of life in terms of its ever more basic elements was, at least in part, deceptive. It did not so much explain 'the mysteries of nature' as displace them to the next level down.

> 'These attempts at the division [of the organism into its component parts] produce many adverse effects,' the German poet and scientific polymath Johann Goethe observed. 'To be sure it is possible to dissect what is alive into its component parts, but from those parts it is impossible to restore it back to life. Each creature is a patterned gradation of one great harmonious whole.'

This might seem no more than the self-evident truism that 'the whole' is vastly more than the sum of its parts – but it is central to the driving underground of that 'vital force'. The project of itemising the several thousand chemical reactions that take place within the cell does not begin to account for those extraordinary properties that so clearly distinguish the living from the non-living, the animate from the inanimate world. Further, the more the intellectual juggernaut of the microscopists and chemists advanced, atomising the living organism into its ever more

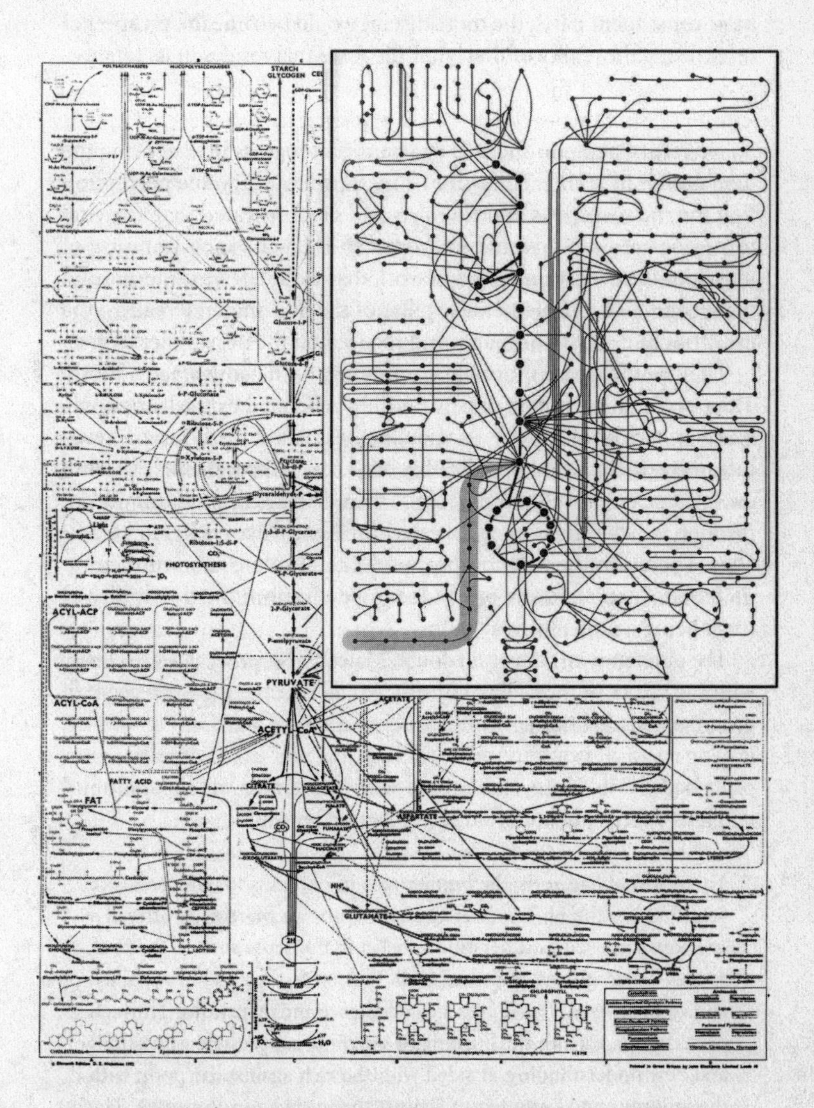

Fig 9-4: *The chemistry of life. Thousands of enzymes within the cell facilitate the transformation of food into the body's 'flesh and blood'. Note, at the centre of this maze of chemical reactions (simplified inset), the circular combustion of the fuel glucose, fizzing like a perpetual Catherine wheel to generate the sparks of energy that animate the body.*

basic constituent parts, the more distant would become the prospect of recognising the reality of that 'vital' life force that resides in its totality.

So we return again to the role of scientific progress in obscuring the dual nature of reality. In the mid-nineteenth century, the perception that the microscopists and chemists had demonstrated that 'the vital force does not exist' would link up with Darwin's incorporation of man into the evolutionary framework that would drive underground the second and complementary pillar of the non-material realm – the exceptionality of the human mind focused on the non-material soul.

They would prove mutually reinforcing, to the advantage of both. The never-ending stream of new findings generated by the laboratory-based science bolstered, by association, the credibility of Darwin's evolutionary theory. Meanwhile, by way of reciprocation, Darwin's evolutionary 'reason for everything' offered an intellectual framework through which scientists could assert their exclusive claims to knowledge. There could be no compromise. The 'absolute' commitment to materialist explanations precluded, by definition, the possibility of there being a non-material realm.

The outcome was scarcely in doubt. Materialist science, allied so closely with the values of the Enlightenment, would become the main agent of progress, rescuing humanity from a dismal past of want and ignorance, leading it to a glorious future. By contrast, the non-material realm, compromised by its close association with the tenets of 'superstitious' religious belief, could offer no such practical benefits.

> The spiritual had sunk the human race in darkness for thousands of years,' writes the philosopher Graham Dunstan Martin. 'It offered a spurious knowledge, superstitious belief in nonsense such as the soul, angels, devils, ghosts, gods and God. It offered stasis and immobility, and ignorant credulity. It preached obedience and incuriosity. Through persecutions, torture and burnings it sought to suppress new questions and new understanding. It sided with the rich against the poor, with the ignorant and complacent against the lively and inquisitive. The moral was clear: reject the spiritual. Matter is all.

Together, Darwin's evolutionary theory and the biologists' reductionist programme would extend the influence of science far beyond its legitimate concerns in explaining 'the phenomena of life', to drive forward the progressive secularisation of Western society. The seemingly self-evident veracity of Darwin's theory would itself become an article of faith for all who aspired to that better future where man would no longer be, in the words of the free-thinking nineteenth-century German physician Ludwig Büchner, 'the humble and submissive slave of a supernatural master, nor the helpless toy in the hands of heavenly powers – but a proud and free son of nature, understanding its laws and knowing how to tutor them to his own use'.

Certainly some scientists, at least initially, proved resistant. But the commitment to Darwin's materialist explanation of the living world would, in time, become a qualification requirement for all who aspired to pursue a career in biology – where to express doubt (at least publicly) was tantamount to confessing to being of unsound (or at least unscientific) mind.

Throughout the nineteenth century and the early part of the twentieth, the remorseless drive of biology's 'reductionist' programme of reducing 'the whole' to its constituent parts accelerated, with the discovery of the autonomic nervous system and a dozen different hormones, numerous vitamins and micro nutrients, the gut enzymes involved in digesting and absorbing food, the diverse ways by which the kidneys filter the blood to remove its waste products, the mechanisms of adaptation to heat and cold, and the specialisation of distinct parts of the brain to fulfil the diverse functions of seeing and hearing, language and movement. By the late 1930s the entire jigsaw of human physiology was nearly complete, and just when it seemed there was nowhere further to go, the massive increase in magnifying power of the recently discovered electron microscope would allow biologists to cross the final frontier, penetrating deep into the astonishing structure and function of the cell to unravel those mysteries of genetic inheritance, protein synthesis and much else beside which would preoccupy them for the subsequent sixty years – before reaching the final destination, the 'instructions' for the secret of life strung out along the Double Helix.

There has never been anything quite like it, the most protracted and sustained project in the history of science. Amidst all the excitement when 'every day brings new discoveries', and inspired by the vision of finding the Holy Grail of the secret of life in the elegant spirals of the Double Helix, it was easy to overlook the profounder questions about the phenomena of life – and particularly how its properties so far surpass those of the most sophisticated of manmade machines, which neither grow from fertilised eggs nor reproduce their kind, neither rebuild their own parts when destroyed, nor maintain their form and structure unchanged while their material composition is in a constant state of flux.

> 'I was present at a meeting of one of the most distinguished of biochemistry departments,' recalls biologist William Beck of a most revealing incident. 'The discussion was interrupted by a message from a local philosophical society asking if the department would provide a speaker to participate in a forthcoming symposium on the nature of life. All assembled understood biochemistry and heredity and genes and enzymes, but no one felt he had anything to say about life. The request was politely declined.'

So too the human mind, and what *it* might be, is conspicuous only by its absence from the discourse of neuroscientists. 'Nobody has the slightest idea how anything material [such as the brain] could be conscious,' observes the cognitive scientist Jerry Fodor. 'Nobody even knows what it would be like to have the *slightest idea* about how anything material could be conscious.'

And that, to put it simply, is how the combination of a progressive evolutionary doctrine and the progressive unravelling of the complexities of biology drove underground the twin pillars of the 'non-material', the formative vital force of life and the human soul. Together they successfully extinguished the challenge to their exclusive claims to knowledge by first conflating the non-material realm with religious belief rooted in the past, while simultaneously allying their exclusively materialist explanations with the Enlightenment vision of a better world. That dual nature of reality, the material and non-material, would remain buried so long as science continued to advance towards that better future. But when its remorseless progress reached its final destination, the genome of man and fly spelt out, the brain observed 'in action' smelling a rose and constructing a sentence, the reality of the non-material domain would, inevitably, re-emerge from the shadows.

For biology, and science generally, this is now both 'the best of times and the worst of times'. The best of times because its prestige has never been greater, the scale of its research institutions never more impressive, its financial support never more lavish. And yet this is also the worst of times, for by the time it is possible to say 'This is how it happened' – this is how the universe and our solar system came into being, this is how the landscape of our earth was formed, this is the universal code of life, and so on – then 'the future', what comes after, is likely to be something of an anticlimax.

The perception that science's glory days may be drawing to a close extends to the practical applied disciplines of technology and medicine on which its reputation as the agent of human progress so soundly rests. The future prospects of the space programme, which so epitomised the adventurous scientific spirit, are constrained by the realisation that we will never travel beyond our solar system to discover other worlds like our own. So too the electronics revolution, where the capacity of the worldwide web and satellite television to communicate information and entertainment far exceeds that of the human mind to absorb it. When it is possible to flick through a choice of seventy channels from the comfort of the sitting-room couch, further technical refinements that might give access to yet more channels would seem scarcely necessary. To be sure, there may be considerable scope for improving their content and quality, but that is a different, non-scientific matter.

So too medicine, where the tidal wave of therapeutic innovation of the post-war years, from heart transplants to test tube babies, provided the most dramatic evidence of the magnificent power of science to free people from the fear of illness and untimely death. But when, as now, most in the Western world live out their natural lifespan to succumb from illnesses strongly determined by ageing, then realistically much of what is do-able has been done.

There is more than a glimpse here of the observations of the main character in Thomas Mann's great novel *Buddenbrooks* (1901):

> I know from life and from history something you have not thought of: often the outward, visible, material signs and symbols of happiness and success only show themselves when the process of decline has already set in. The outer manifestations take time – like the light of that star up there which may in reality already be quenched when it looks to us to be shining at its brightest.

The light of those stars we now see shining brightly in the heavens (or relayed back from the Hubble telescope) has taken millions of years to reach us, by which time the energy from which it flows has long since been exhausted. Similarly, the light of scientific success that now shines so brightly was generated by the endeavours of many, stretching back over the past four hundred years. Where now are those new ideas, the fresh fruits of research and innovation to maintain that momentum?

For the best part of twenty years the prospects of further advance have rested almost exclusively on the shoulders of the New Genetics and neuroscience, whose potent, and now technically routine, methods for generating new knowledge have swept all other forms of investigation aside, to become virtually synonymous with the scientific enterprise itself. To the uncritical eye these generously funded bastions of Big Science could not be more productive, filling the pages of academic journals with new findings on an unprecedented scale – a year's worth of (for example) the *Journal of Biological Chemistry* now fills an entire library shelf.

But the fundamental premise no longer holds that the New Genetics and neuroscience would make clear all that is currently obscure. They have certainly, as we have seen, revolutionised our understanding of genetic inheritance and the workings of the brain – though not in the way intended. They have revealed, too, much of interest with practical application in both medicine and agriculture, such as the discovery of the genes involved in specific diseases and the genetically engineered version of human insulin. But in general the fortunes of the biotechnology industry that sought to turn those discoveries to commercial advantage have been disastrous, 'one of the biggest money-losing industries in the history of mankind', the Chief Executive of Genentech informed financial analysts in New York in 2006 – to the tune of $100 billion over a period of twenty years. There is a strong impression, too, of sensationalism in the extravagant claims made on behalf of cloning, experiments on human embryos and similarly controversial cutting-edge fields of research whose ability to generate publicity far outstrips any evidence of practical benefit.

The New Genetics and neuroscience have both proved to be, as it were, one step too far in biology's 'reductionist' programme. Biologists now find themselves in the invidious situation of having to suppose that the sheer accumulation of yet more facts about 'which genes do what' might somehow, like a bulldozer, drive a causeway through the perplexities posed by, for example, those 'master genes' –

without the slightest inkling of a theory of what their accumulated facts might all add up to. And so too neuroscientists and the cardinal mysteries of the mind. The overwhelming impression is of labourers excavating a vast hole, where the more energetically they dig, the deeper it becomes. Or, to switch analogies, 'Under the Upas tree nothing grows,' and the continuing dominance of the Big Science of the New Genetics and neuroscience, with their capacity to generate an ever vaster accumulation of facts, yet more genome projects and brain-imaging studies, has stifled the 'spirit of science', and with it the investigation of the anomalies in current explanations on which genuine scientific progress depends. They have become 'degenerate research programmes', as described by philosopher Herbert Dreyfus of the University of California. Such a programme

> starts out with great promise, offering an approach that leads to impressive results in a limited domain. Almost inevitably researchers will want to try to apply the approach more broadly . . . As long as it succeeds the research programme expands and attracts followers. If, however, researchers start encountering unexpected but important phenomena *that consistently resist the new techniques* the programme will stagnate and researchers will abandon it as *an alternative approach becomes available.*

The prospects for that 'alternative approach', and its fruitful consequences, are perhaps more immediate than might be supposed. Still, for the moment, the merry-go-round of life goes on, and the failure to find the secret of life in the Double Helix or to 'crack' the brain makes not a jot of difference to people's lives. The confounding of Darwin's evolutionary theory similarly is of no practical significance. The near-universal belief prevails, at least in the West, that Darwin solved the problem of man's origins, and indeed his evolutionary theory functions well enough as the great cosmogenic myth of the twentieth century, just as did the cosmogenic myths of Genesis that preceded it. It is the nature of our species to pose such questions, and equally in the absence (by definition) of any certain answers, to construct our myths accordingly.

> 'Theories of evolution reinforce the value system of their creators by reflecting their image of themselves and of the society in which they live,' observed the Oxford historian John Durant at a meeting of the

> British Association for the Advancement of Science in the early 1980s. 'Time and again ideas of human origins turn out on closer examination to tell us as much about the present as about the past, as much about our own experiences as those of our remote ancestors.'

Many no doubt will prefer to stick to the 'scientific version' of our origins, and indeed look favourably on its substantial contribution to the progressive secularisation of Western society. But it cannot be that the current state of science, that has played so powerful a role in framing the modern mind, is of no concern. The general rule holds that the uncritical endorsement of misleading explanations can have grievous consequences. We have glimpsed in an earlier chapter some of those in the propagation of eugenic policies and the absurdities of sociobiology. But there is more, for, paradoxically, despite 150 years of remorseless scientific progress, we are left with a surprisingly pessimistic view of humanity as the perpetrators of the terrible destructive wars of the past century and the destroyers of the planet that sustains us. 'We eat well, we drink well, we live well,' comments the cultural historian Michael Ignatieff, 'but we no longer have good dreams.'

The many reasons for that pessimism become clearer when we contrast the present age with that most vibrantly optimistic of times, the eighteenth-century Enlightenment whose values pervaded western Europe – '*illuminazione*' in Italian, '*lumière*' in French, '*Aufklärung*' in German. Its catchphrase, the German philosopher Immanuel Kant suggested in an essay written in 1784, should be '*sapere aude*' – 'dare to know': not just scientific knowledge but self-knowledge, the knowledge that would set man free to create a society based on tolerance and freedom. The citizens of eighteenth-century Europe enjoyed none of the benefits of modern science, their lives compared to ours were plagued by physical suffering, but unlike us they had a vision of a better world, set to music by Bach and Handel, Mozart, Haydn and Beethoven. That optimism was underpinned by the recognition of man's exceptionality, the desire, as the writer Kenan Malik puts it, 'to place Man at the centre of philosophical debate, to glorify his abilities and to see human reason as the tool through which to understand nature and himself'.

Now, as then, people lead purposeful and fulfilled lives – the more so, one would suppose, for being the beneficiaries of all the wonders of modern technology. But the vibrant optimism of that Enlightenment view of man has evaporated. Indeed, we could almost be said to live in an age of 'counter' enlightenment – where the prevailing scientific view maintains that man's sense of himself as an autonomous independent being is no more than an illusion generated by his brain, and the joys and agonies of human love no more than a device for the propagation of his genes.

Many factors have contributed to that cultural pessimism, but the most obvious feature that distinguishes modern man from Voltaire and his contemporaries is the ascendancy of scientific materialism, and the loss of an appreciation of there being a non-material reality that transcends our everyday concerns. We have lost that sense of living in an enchanted world. We might now, thanks to science, comprehend the universe of which we are a part, only to discover that its properties, as evolutionary biologist Richard Dawkins puts it, 'are precisely those we should expect if there is, at bottom, no design, no purpose, no evil and no good – nothing but blind, pitiless indifference'. We have lost, too, sight of the most significant factor of all – the exceptionality of the human mind. We no longer appreciate what at one time seemed self-evident: the extraordinariness of possessing a mind whose powers of reason can distinguish truth from falsehood. There is nothing remotely scientific in denying the most certain thing we know, our sense of self; but 'On the maps provided by contemporary science,' writes Bryan Appleyard, 'we find everything except ourselves.'

We have experienced, in short, a drastic narrowing of our horizons and our appreciation of what might be possible. 'As for the meaning of life, I do not believe it has any – and it is a source of great comfort,' observed the philosopher Isaiah Berlin. 'We make of it what we can and that is all there is to it. Those who seek for some cosmic all-encompassing explanation are deeply mistaken.'

There is, of course, a bleak heroism in this view of man, who, having shed all the illusions that enriched his life in the past, now recognises he is no more than the fortuitous consequence of impersonal natural laws, an unusual life form on an insignificant planet in a remote corner of the universe. But the findings of the recent past have changed all that. We can no longer be certain, as we might have been till recently, that Professor Berlin is right in supposing 'that is all there is to it', because we cannot, by definition, comprehend the nature of that potent non-material realm with its power to conjure the wonders of life from the bare material bones of scientific knowledge.

It is only by recognising the narrow confines of the materialist view that it becomes possible to move on, comments the philosopher Roger Scruton, 'to replace the sarcasm which knows that we are merely animals, with the irony that sees that we are not'. The time has come to break the silence and restore a coherent, balanced view of ourselves and our world by putting aside biology's foundational evolutionary theory and embracing the dual nature of reality. This may perhaps seem highly improbable, but it cannot be long before a proper appreciation of the true significance of the findings of the recent past begins to sow doubts in inquisitive minds.

10

Restoring Man to his Pedestal

'Dare to know!'

Immanuel Kant

We turn now to the future, perhaps ten years from now, to imagine how differently our world might appear once the astonishing discoveries of the recent past have become common knowledge. The great drawback of Darwin's simple, all-encompassing evolutionary theory has always been that it robs the living world of its unknowable profundity. There is nothing too wonderful, awe-inspiring or bizarre that it cannot account for as having evolved to be so over millions of years. When the same ready explanation is applied to each and every fact of biology – no matter how extraordinary – the phenomena of life are reduced to mere banalities.

Strip away the comfort blanket of Darwin's Reason for Everything and the 'facts' remain the same, but now appear fresh-minted in glorious Technicolor as astounding and amazing, magical and mysterious: the automated factory of the cell, the Arctic tern's twenty-five-thousand-mile migration, the geometrical design of a bat's face, the surprisingly non-sloth-like toilet habits of the giant sloth – every detail, in short, of the billionfold complexities of living organisms. The most obvious of facts appear distinctly non-obvious. How to begin to account for the sheer profligacy of life, whose limitless novelties of form encompass the entire range of what might be possible, from the elephant's trunk and giraffe's neck to the 'long-nosed bugs, luminous beetles, harmless butterflies disguised as wasps, wasps shaped like ants, sticks that walk, leaves that open their wings and fly' of the Amazonian forest? And so on, *ad infinitum*, for any list of such novelties would end up itemising virtually every significant characteristic of every living thing on earth.

The same sense of astonishment applies in equal measure to all those critical innovations that mark the stages of the history of life as we recognise how, in truth, we know virtually nothing about how they came to be: the first single-celled organisms; the genetic code; the photosynthesis by which plants capture the energy of the sun to drive the great cycle of life; the first multi-cellular life forms of the Cambrian explosion, aware through their senses of sight and touch of the world around them; the mechanism of those transitions from fish to reptile, to mammals, to birds, each stage initiating a further 'explosion' of millions of new and unique species.

The setting aside of that evolutionary doctrine would also drag from the shadows all those very important facts passed over or censored in the conventional account lest they raise doubts as to its veracity – none more so than the riddle of the origins of our species, epitomised by the hazards of standing upright and acquiring that prodigiously enlarged brain, complete with its 'language acquisition device'. That censorship of significant facts extends, most obviously, to the human mind, where the self-evident truism that intellectually we share (as we must) much in common with our primate cousins is a triviality when set against the manifold differences so long concealed. Now, once again, we can begin to appreciate those distinct properties that so unambiguously set us apart. We have access, through the faculty of language, to thoughts and ideas, our own and those of others. Our primate cousins do not. We possess the power of reason, through which we comprehend causes and their effects and can thus make sense of the world around us. They do not. We can imagine things to be different from how they are, and plan for our futures. They cannot. We know our beginnings and our end, and recognising the fact of our mortality, are impelled to seek explanations for our brief sojourn on earth. They do not. We inhabit the spiritual domain centred on the self, the soul, the 'I', with its several distinct interconnected parts which, being non-material, and thus not constrained by the material laws governing the workings of the brain, is free to choose one thought over another or one course of action over another. And that inextricable connection between the non-material self and *freedom* is the defining feature of man's exceptionality, for we, unlike our primate cousins, are *free* to forge our own destinies to become that distinct, unique person responsible for our actions of which all human societies are composed, and from which virtually everything we value flows.

Science can know nothing of this, and thus, by necessity, denies man his freedom and the reality of his soul, with all the grievous consequences already touched on. The philosophic view alone recognises the dual nature of reality, and reasserting its priority would prompt one of those rare convulsive upheavals in human understanding that the historian Thomas Kuhn described as a 'paradigm shift' from one way of 'knowing' to another. It is easy to suppose that we are on a never-ending upward trajectory of knowledge, as we know so much more about the natural world than a century ago, and more still than the century before that. But, as Kuhn pointed out, it is not like that. Science is for the most part a humdrum problem-solving affair, filling in the details of the broad picture of what is already known while the anomalies and inconsistencies that do not fit prevailing theories are readily overlooked. There comes a time, however, when cumulatively those anomalies can no longer be ignored, at which point science enters into 'a state of crisis' which can only be resolved by a radical shift in its fundamental theories and its perception of 'how things are'. These paradigm shifts include most obviously Galileo's liberation of astronomy in the seventeenth century from the certainty of an earth-centred cosmos, relocating it as just one of several planets in orbit around the sun. The subsequent massive expansion in intellectual horizons is the defining characteristic of a 'paradigm shift' in revealing how little the prevailing theories explain, while simultaneously opening the door to knowing so much vastly more than was previously thought possible.

We are, it seems, on the brink of a similar paradigm shift, not once but twice over, in our understanding of both the natural world and of ourselves. We are led through the New Genetics to the necessity to set aside Darwin's evolutionary doctrine. It has served its purpose well enough as a tidy, all-encompassing 'reason for everything', providing the wider context within which biologists could pursue their investigations without fear that the processes so revealed might defy any simple materialist explanation. But as time has passed its anomalies and inconsistencies have proved ever more pressing, and now, with the crushing verdict of the genome projects, we are left to stare into the abyss of our ignorance of virtually every aspect of the complexities of the living world and its evolutionary history. And yet, conversely, our knowledge of the natural world becomes vastly greater than could be imagined, as we can now recognise those profundities for what they are, rather than seeing them through that distorting prism which would suppose them to be so much simpler than they really are.

Again we are led through the findings of the Decade of the Brain to recognise the unfathomable depths of the workings of its neuronal circuits and their relationship to the human mind. Yet conversely we open the door to knowing so vastly more than we could imagine by restoring our confidence in the intuition of our extraordinary minds into the reality of the non-material realm. Thus we come to deepen our understanding by recognising how little we know, and through meditating on the profundities touched on in this account: the mysterious creative evolutionary force which from the beginning has conjured ever more complex forms of organisation from the simplest elements of matter; the inscrutable origins of that cell with its capacity to bring into being every life form that has ever existed; the sudden, dramatic emergence of new forms of life from the Cambrian explosion onwards; those seemingly mundane practicalities of our everyday lives of the Introduction: seeing, talking, growing and reproducing our kind; the phenomenal information-processing power of the human brain; the potency of the forces of the non-material realm – and the (rediscovered) human soul.

The major beneficiary of this new understanding will be science itself, liberated from the dead hand of evolutionary certainty and the ever deepening hole of Big Science's degenerate research programmes to turn its attention to all those substantial questions which have so far eluded its methods of investigation. How do birds navigate their thousand-mile journeys across the featureless ocean? Where, in its depths, are the breeding grounds of whales? What happens to the chemistry of the brain to cause us to sleep and dream, and why should sleep refresh our bodies? What events deep in the earth cause earthquakes to quake and volcanoes to erupt?

Biologists, freed from the obligation to subscribe to the evolutionary doctrine, could turn their attention to those many observations and anomalies that contradict its too ready explanations. 'All nature is at war, one organism with another,' claimed Darwin – but it is not so, for the most striking feature of the natural world is not the competitive struggle for existence, but its antithesis – cooperation. So while in the past biologists have interpreted much that is apparently cruel and wasteful in the living world as evidence of a random, purposeless evolutionary process, the greater challenge is to account for the much greater evidence of harmony and purpose.

There is not a single species that does not have a mutually beneficial relationship with another. That cooperation takes many forms, most obviously in providing a hospitable, if sometimes surprising, refuge, such as a species of crab that lives within the rectum of the sea urchin, or another within the shell of the oyster. The sea worm *Urechis caupo* is nicknamed 'the innkeeper' for giving space to thirteen species of fish, arthropod and mollusc in the u-shaped burrow it makes in the Californian mudflats. Mutual cooperation is evident too in the transport that mobile creatures provide for stationary ones, and the manner by which bees, honey birds and bats are all rewarded by plants with nutritious nectar for facilitating their cross-fertilisation. Then there are numerous examples of mutual protection: where, for example, baboons associate with gazelles on the plains of Africa and profit from their keen sense of smell, while the gazelles benefit from the baboons' superior visual acuity in detecting predators. Many species foster the health of others by providing a cleaning service, as in the blowfly's preference for laying its eggs in the festering wounds of animals. 'This might appear to be one of nature's great cruelties,' observes George Stanciu of the University of Michigan, 'but when the larvae hatch they feed on the pus and consume dead tissues, even their excretions disinfect the wound! Far from being cruel, the fly larvae may be the animal's only chance to recover from a possibly lethal infection.'

A greater challenge still would be to pursue the many neglected observations and experiments over the past hundred years that hint at the existence of yet unknown forces in biology, such as that remarkable capacity of living organisms to regenerate their parts. This presupposes, as biologist Rupert Sheldrake points out, the existence of some 'field' by which an organism knows itself, and its parts, in their entirety. How else to account for the facility with which limbs, once broken, heal and remodel themselves as before, or the capacity of the brain to compensate for blindness by massively increasing the acuity of its other sense organs, such as touch and hearing? How otherwise, too, could the humble flatworm when cut in two reconstruct itself as two separate organisms, the head acquiring a new tail, and the tail a new head? There is more than a hint too of those hidden biological forces in the evocative beauty and harmony of the geometric laws of proportion of the human body, whose several parts – head, limbs and thorax –grow in synchrony with each other and with the body as a whole. That synchronous growth from embryo to adulthood reflects how, as the biologist D'Arcy Wentworth Thompson puts it, 'throughout its fabric one part is related and fitted to another'.

The reinvigoration of science with novel (or rediscovered) ideas and better theories might in turn be expected to counter the current disillusion with a subject that for many seems no more than a disconnected (and often poorly taught) collection of tedious facts, divorced of any grand vision of what it might all add up to. It is possible to imagine that the recent and precipitous fall in the numbers of science graduates might be reversed were the subject reinfused with that recognition of the unfathomable profundities of the natural world that for Isaac Newton, likening himself to 'a boy playing on the seashore', was so powerful an incentive to 'wonder why'.

The new paradigm must also lead to a renewed interest in and sympathy for religion in its broadest sense, as a means of expressing wonder at the '*mysterium tremendum et fascinans*' of the natural world. It is not the least of the ironies of the New Genetics and the Decade of the Brain that they have vindicated the two main impetuses to religious belief – the non-material reality of the human soul and the beauty and diversity of the living world – while confounding the principal tenets of materialism: that Darwin's 'reason for everything' explains the natural world and our origins, and that life can be 'reduced' to the chemical genes, the mind to the physical brain.

There is, of course, nothing in the new paradigm that can be interpreted as direct evidence for a Creator, or that would resolve the insuperable difficulty for many of conceiving of his existence and purpose. But while it is as hard as ever to imagine him hard at work designing several thousand species of beetle, there is vastly greater evidence of 'design' – for those who would wish to interpret it as such – than the supposition that the vast panoply of nature should be the incidental consequence of those numerous random genetic mutations that the genome projects have so unequivocally failed to identify. Indeed, those two critical innovations, the genetic code and language, might almost seem to *require* a higher intelligence to devise the code and draw up those rules of syntax. There is, after all, nothing intrinsic in the chemical molecules strung out along the Double Helix to explain why they should carry a specific genetic message. Or, put another way, the message is 'extraneous' to the chemical and physical properties of the molecules themselves. So, just as it requires human intelligence to produce anything with a high information content, whether books or dictionaries, music scores or compact discs, so by analogy it would be reasonable to infer that it would require a 'higher intelligence' to formulate the genetic code.

This puts a rather different gloss on the influential arguments of the main protagonists of the evolutionary doctrine, Richard Dawkins, Daniel Dennett and Ed Wilson, who seek to explain away the near-universality of some form of religious belief as a self-gratifying delusion to make us feel better about the vagaries of our lives. Their remorseless hostility to religion ('a dangerous collective delusion') is best interpreted as a rhetorical device, a sleight of hand to distract attention from the intellectual weaknesses of scientific materialism. Dawkins, in a typical passage, invokes the standard juxtaposition of rational science and irrational religion to buttress his assertion that faith is 'one of the world's great evils':

> 'It is fashionable to wax apocalyptic about the threats to humanity,' he writes, 'but I think a case can be made that faith, the principal vice of any religion, is one of the world's great evils, comparable to the smallpox virus but harder to eradicate. Faith is a great cop-out, the great excuse to evade the need to think and evaluate evidence.'

There is, to be sure, ample evidence for the wickedness carried out in religion's name that is the grist to Professor Dawkins' influential argument. But one could as well, in the light of the findings of the recent past, reverse his juxtaposition as being one rather between *rational* faith and *irrational* science. For many, such as Voltaire, it has seemed reasonable, both intellectually and intuitively, to infer from the wonders of the natural world and the exceptionality of the human mind the existence of some Divine Being. From this perspective, faith is not, as Dawkins claims, 'the great excuse to evade the need to think and evaluate evidence', but on the contrary, as the writer C.S. Lewis describes it, 'the art of holding on to things your reason has once accepted'.

By contrast, it now seems deeply irrational for materialist science to deny the exceptionality of the human mind and to insist that the sense of self and free will are no more than illusions generated by the workings of the brain. It is certainly irrational to assert the truth of the evolutionary doctrine ('the mystery of our existence is a mystery no longer because Darwin solved it') in the face of all the scientific evidence that would contradict it.

This renewed sympathy with religion in keeping alight the flame of 'rational faith' against the rising tide of secularism would also, it might be anticipated, heal that rupture in Western civilisation between its present and its overwhelmingly Christian past, whose beliefs and traditions are an expression of that assumption, common to all the major religions, of there being 'more than can be known'. It is scarcely possible to properly appreciate the unparalleled riches of Western culture without recognising the profundity of the religious beliefs that inspired the sublime music of eighteenth-century composers like Bach and Handel, or great painters like Raphael and Michelangelo, or the anonymous builders who created the great cathedrals of Christendom. It would be good, too, to think that future generations might have a better understanding of those philosophers and thinkers stretching back to Plato, whose reflections on the meaning and purpose of human existence and profound insight into the reality of the human soul we have only belatedly rediscovered. It would be good, too, to think that the school curriculum might in future imbue in young minds a proper appreciation of the brilliant achievements of that first civilisation of Cromagnon man who 'worked it all out for himself' – and the deep riddle of his evolutionary origins.

And it might be hoped that a proper appreciation of our exceptionality might restore that sense of gratitude for the 'natural miracle' of the world around us: the bounteous and unexplained plenitude of the fish in the sea, the birds in the air and the fruits of the orchard in all their infinite (and unfathomable) diversity. And encourage too, along the way, a deepened respect for the planet that sustains us that might guarantee its future, and our own, in the troubled times ahead.

What then to make of Charles Darwin, who has cast so long a shadow over the past 150 years? He was, like so many of his contemporaries in that Golden Age of Natural History, a brilliant naturalist in extraordinary times. He had the audacity to seek a grand unifying explanation, in the tradition of his fellow countryman Isaac Newton, for the processes of life and its history. But those processes, so many billion-fold times more complex than the laws of gravity, defy such simplification. His legacy then is rather different from that commonly perceived. Together with Marx and Freud, he is one of that triumvirate of imaginative thinkers of the nineteenth and early twentieth centuries whose assertion of the priority of the scientific view 'would occupy the centre stage of Western thought for so long'.

> 'Darwin made theological and spiritual explanations of life superfluous,' Douglas Futuyma of the University of New York State observes. 'Together with Marx's theory of history and Freud's attribution of human behaviour to influences over which we have little control, the theory of evolution was a crucial plank in the platform of materialist science.'

Each was inspired by an important insight. Marx was led through his concern for the plight of the industrial working class to formulate a penetrating analysis of capitalist society. Freud perceived the formative influence of a child's instinctive desires. Darwin recognised the differences between closely related species, and suggested they were not immutable, but could 'evolve one into another'.

Each then elaborated those insights into a universal theory. There is no feature of society, its laws, institutions and culture that Marxists could not interpret as a mechanism for 'capitalist exploitation'. Freud's famously all-encompassing explanations are illustrated by the story of the man who visited his analyst: if he was early, the analyst inferred he was anxious; if he was on time, he was compulsive; if he was late, he was resentful. Darwin's 'reason for everything' has been (more than) touched on.

Each denied the reality of the self as an autonomous being 'free to choose', claiming rather that human action is determined by a powerful hidden force, whether economic, psychic or (loosely speaking) genetic – over which the individual has no control. Each saw himself as a materialist, and was hostile to religion, which for Marx was famously 'the opium of the people', and for Freud 'an infantile yearning for a powerful protective father', while Darwin confided in his cousin Francis Galton, 'I gave up religious belief almost independently of my own reflections.'

Darwin's contributing 'plank' to that 'platform of materialist science' alone endures – but for how much longer? It certainly seems surprising in retrospect that Marx's and Freud's self-evidently erroneous theories should have proved so persuasive to so many and for so long. Now it is the turn of Darwin, whose reputation can scarcely survive the devastating verdict of the findings of the recent past. Before long he must fill that vacant chair in heaven alongside Marx and Freud, at which point the triumvirate will be complete. As for ourselves, the eighteenth-century poet Alexander Pope could not, it now appears, have put it more succinctly:

> Sole judge of Truth, in endless Error hurled;
> the glory, jest and riddle of the world!

Acknowledgements

The protracted gestation of this book has been more than rewarded by the opportunity to learn from and explore the insights and ideas of those who personally or through their writings have shaped its arguments. They include: Denis Alexander, Bryan Appleyard, Robert Augros, Michael Behe, Peter Bowler, Sean Carroll, Simon Conway Morris, John Cornwell, Michael Denton, Robert Foley, Henry Gee, Brian Goodwin, John Haldane, Gertrude Himmelfarb, John Horgan, Christopher Howse, Phillip Johnson, Richard Lewontin, John McCrone, Colin McGinn, Alister McGrath, Kenan Malik, Graham Dunston Martin, Robert Matthews, Mary Midgley, Michael Morange, Thomas Nagel, Anthony O'Hear, John Polkinghorne, Andrew Pomiankowski, Matt Ridley, Steven Rose, Jeffrey Schwartz, Roger Scruton, Rupert Sheldrake, George Stanciu, Mikael Stenmark, Raymond Tallis, William Uttal, Keith Ward, Jonathan Wells and Robert Wesson.

My heartfelt thanks to the ever-helpful librarians at the Royal Society of Medicine, the London Library and the Wellcome Institute; to publishers Michael Fishwick, Richard Johnson and Arabella Pike in the UK and Dan Frank in the United States; to my editor Robert Lacey and agents Carolyn Dawnay, Marcella Edwards and Emma Parry; to Vanessa Adams for her prodigious secretarial skills and (literally) invaluable support; and most of all to those who have been there steadfastly from the beginning, my wife Juliet for her wise counsel, and my children Frederick and Allegra.

Notes

Chapter 1: Science Triumphant, Almost

Page 1 – We live in the Age of Science . . .
The most comprehensive and accessible account of the scientific achievement of the past sixty years is Neil Schlager (ed), *Science and its Times: Understanding the Social Significance of Scientific Discovery*, vol 7 (Gale Group, 2000). See also Bryan Bunch and Alexander Hellemans, *The History of Science and Technology* (Houghton Mifflin, 2004); John Gribbin, *Science: A History, 1543–2001* (Allen Lane, 2002); Ivan Amato (ed), *Science Pathways of Discovery* (John Wiley, 2002)
Page 2 – 'a moment of glory too swift. . .'
Bill Bryson, *A Short History of Nearly Everything* (Black Swan, 2004)
Page 3 – 'Tom [Gray] and I had surveyed. . .'
Donald C Johanson and Maitland A Edey, *Lucy: The Beginnings of Humankind* (Penguin, 1990)
Page 6 – The Double Helix. . .
The classic history of the events leading up to the discovery of the Double Helix is Horace Freeland Judson, *The Eighth Day of Creation: Makers of the Revolution in Biology* (Penguin, 1979). See also Robert Dolby, *The Path to the Double Helix: The Discovery of DNA* (Dover Publications, 1974); Michael Morange, *A History of Molecular Biology* (Harvard University Press, 1998); Lily E Kay, *Who Wrote the Book of Life?: A History of the Genetic Code* (Stanford University Press, 2000); James Watson, *The Double Helix* (Weidenfeld & Nicolson, 1968)
Page 6 – This situation would change dramatically. . .
The science behind the three technical innovations is described in James Le Fanu, *The Rise and Fall of Modern Medicine* (Little, Brown, 1999)
Page 7 – It lies beyond hyperbole. . .
David E Comings, 'Prenatal Diagnosis and the "The New Genetics" ', *American Journal of Human Genetics*, 1980, vol 32, p 453
Page 8 – Thus the Human Genome Project (HGP) was born. . .
Victor McKusick, 'Mapping and Sequencing the Human Genome', *New England Journal of Medicine*, 1989, vol 320, pp 910–16; see also Christopher Wills, *Exons,*

Introns and Talking Genes: The Science Behind the Human Genome Project (Basic Books, 1991); Richard Lewontin, 'The Dream of the Human Genome', *New York Review of Books*, 28 May 1992

Page 8 – 'like a mechanical army. . .'

John Savile, 'Prospecting for Gold in the Human Genome', *BMJ*, 1997, vol 314, pp 43–9

Page 9 – Meanwhile, the human brain too. . .

The best account of modern neuroscience, on which this account draws extensively, is John McCrone, *Going Inside: A Tour Around a Single Moment of Consciousness* (Faber & Faber, 1999)

Page 10 – Here again, a series of technical innovations. . .

Marcus E Raichle, 'Behind the Scenes of Functional Brain Imaging: A Historical and Physiological Perspective', *Proc Nat Acad Sci*, 1998, vol 95, pp 765–72; Marcus E Raichle, 'Visualising the Mind', *Scientific American*, April 1994, pp 36–42

Page 11 – The completion of the first draft of the Human Genome Project. . .

Kevin Davies, *Cracking the Genome: Inside the Race to Unlock Human DNA* (Johns Hopkins University Press, 2001); see also John Sulston and Georgina Ferry, *The Common Thread: A History of Science, Politics, Ethics and the Human Genome* (Bantam Press, 2002); Nicholas Wade, *Life Script* (Simon & Schuster, 2001)

Page 11 – The following year, in February 2001. . .

The two papers reporting the findings of the Human Genome Project are 'International Human Genome Sequencing Consortium', *Nature*, 2001, vol 409, pp 860–921 and J Craig Venter, 'The Sequence of the Human Genome', *Science*, 2001, vol 291, pp 1304–49. See also G Subramanian et al., 'Implications of the Human Genome for Understanding Human Biology and Medicine', *JAMA*, 2001, vol 286, pp 2296–307; Carina Dennis and Richard Gallagher, *The Human Genome* (Palgrave, 2001)

Page 13 – The goals of the Decade of the Brain. . .

Eleven essays surveying the full range of neuroscience during the Decade of the Brain written by many of its leading players can be found in *Scientific American*, September 1992. See also Rita Carter, *Mapping the Mind* (Weidenfeld & Nicolson, 1998); Michael I Posner and Marcus Raichle, *Images of the Mind* (Scientific American Library, 1994)

Page 13 – 'As surely as the old system . . ..'

Semir Zeki, *A Vision of the Brain* (Blackwell Scientific, 1993)

Page 14 – 'every facet of mind . . ..'

Steven Pinker, 'Will the Mind Figure out how the Brain Works?', *Time*, 10 April 2000, p 90

Page 14 – First, there is the 'numbers problem'. . .

James Randerson, 'Fewer Genes, Better Health', *New Scientist*, 13 July 2002, p 19

Page 15 – But no more so than. . .

'The Chimpanzee Sequencing and Analysis Consortium', *Nature*, 2005, vol 437, pp 69–87

Page 15 – 'The realisation that a few genetic...'
Svante Paabo, 'The Human Genome and Our View of Ourselves', *Science*, 2001, vol 291, p 1219
Page 15 – 'We cannot see in this why...'
Elizabeth Culotta, 'Chimp Genome Catalogues Differences with Humans', *Science*, 2005, vol 309, pp 1468–9
Page 16 – 'One of those rare and wonderful moments...'
Evelyn Fox Keller, *Making Sense of Life* (Harvard University Press, 2002)
Page 16 – And so, too, the Decade of the Brain...
The perplexities of the findings of the Decade of the Brain are explored in John Horgan, *The Undiscovered Mind: How the Brain Defies Explanation* (Weidenfeld & Nicolson, 1999); see also William Uttal, *The New Phrenology: The Limits of Localising Cognitive Processes in the Brain* (MIT Press, 2001); David J Chalmers, 'The Puzzle of Conscious Experience', *Scientific American*, December 1995, pp 62–5
Page 18 – 'This abiding tendency...'
David Hubel, *Eye, Brain and Vision* (Scientific American Library, 1988)
Page 18 – 'Suppose I know everything about your brain...'
Colin McGinn, *The Mysterious Flame* (Basic Books, 1999)
Page 19 – 'We seem as far from...'
John Maddox, 'The Unexplained Science to Come', *Scientific American*, December 1999

Chapter 2: The Ascent of Man: A Riddle in Two Parts

Page 24 – 'Alone in that vastness...'
Jean-Marie Chauvet, *Chauvet Cave* (Thames & Hudson, 1996)
Page 25 – She is the 'Dame de Brassempouy'...
Édouard Piette, *L'Art pendant l'Âge du Renne* (Masson, 1907); see also Henri Delporte (ed), 'La Dame de Brassempouy', *Études et Recherches Archéologiques de l'Université de Liège* (Liège, 1995)
Page 26 – The Cromagnons' arrival in south-western Europe...
The Cromagnon achievement is well described in Ian Tattersall, *Becoming Human* (Oxford University Press, 1998); see also Steven Mithen, *The Prehistory of the Mind* (Phoenix, 1998); Paul Mellars, 'Major Issues in the Emergence of Modern Humans', *Current Anthropology*, 1989, vol 30, pp 349–89; John Wymer, *The Palaeolithic Age* (Croom Helm, 1984)
Page 27 – And they had a passion for art...
Paul G Bahn, *The Cambridge Illustrated History of Prehistoric Art* (Cambridge University Press, 1998); see also Evan Hadingham, *Secrets of the Ice Age: The World of the Cave Artists* (Heinemann, 1980); Paolo Graziosi, *Palaeolithic Art* (Faber & Faber, 1960); Paul G Bahn and Jean Vertut, *Journey Through the Ice*

Age (Seven Dials, 1997); Anna Sieveking, *The Cave Artists* (Thames & Hudson, 1979); David Lewis-Williams, *The Mind in the Cave* (Thames & Hudson, 2002); John Halverson, 'Art for Art's Sake in the Palaeolithic', *Current Anthropology*, 1987, vol 28, pp 63–89; André Leroi-Gourhan, *The Art of Prehistoric Man in Western Europe* (Thames & Hudson, 1968); Randall White, 'Visual Thinking in the Ice Age', *Scientific American*, July 1989, pp 74–81

Page 31 – The common understanding. . .

Charles Darwin, *On the Origin of Species* (John Murray, 1859) and *The Descent of Man* (John Murray, 1871)

Page 32 – Details aside, one single, powerful image. . .

Thomas Huxley, *Evidence as to Man's Place in Nature* (Williams & Norgate, 1863)

Page 33– And so the major archaeological discoveries . . .

Five standard textbooks on human evolution are Robert Boyd and Joan Silk, *How Humans Evolved* (Norton, 2006); Chris Stringer and Peter Andrews, *The Complete World of Human Evolution* (Thames & Hudson, 2005); Roger Lewin and Robert Foley, *Principles of Human Evolution* (Blackwell, 2004); Steve Jones (ed), *The Cambridge Encyclopaedia of Human Evolution* (Cambridge University Press, 1992); Eric Delson et al. (eds), *Encyclopaedia of Human Evolution and Prehistory* (Garland Publishing Inc, 2000); see also Bernard G Campbell, *Human Evolution: An Introduction to Man's Adaptations* (Aldine Publishing Company, Chicago, 1966); Tim Crow (ed), *The Speciation of Modern Homo Sapiens* (Oxford University Press, 2002). The several more popular accounts include Ian Tattersall, *Becoming Human* (Oxford University Press, 1998); Chris Stringer and Robin McKie, *African Exodus: The Origins of Modern Humanity* (Pimlico, 1996); Desmond Morris, *The Naked Ape: A Zoologist's Study of the Human Animal* (Vintage, 1994); Richard Leakey and Roger Lewin, *Origins Reconsidered: In Search of What Makes us Human* (Little, Brown, 1992)

Page 34 – Those two near-complete skeletons were. . .

Donald Johanson and Maitland Edey (Penguin, 1981), op. cit.

Page 35 – Lucy's novel method of locomotion. . .

Richard L Hay and Mary D Leakey, 'The Fossil Footprints of Laetoli', *Scientific American*, February 1982, pp 38–45; Tim White and Gen Suwa, 'Hominid Footprints at Laetoli: Facts and interpretations', *American Journal of Physical Anthropology*, 1987, vol 72, pp 485–514)

Page 35 – 'The footprint records a normal positioning. . .'

John Eccles, *Evolution of the Brain* (Routledge, 1989)

Page 36 – Ten years later, in 1984. . .

Richard Leakey, *The Origin of Humankind* (Phoenix, 1994)

Page 37 – 'In man the most precise function. . .'

John Napier, *The Roots of Mankind* (Allen & Unwin, 1971)

Page 37 – This extra inch of the human thumb. . .

For a brilliant account of the cultural significance of the human hand see Raymond Tallis, *The Hand: A Philosophical Inquiry into Human Being*

(Edinburgh University Press, 2003); see also Eric Trinkaus, 'Evolution of Human Manipulation', in Steve Jones (ed), 1992, op. cit.; Sherwood Washburn, 'Tools in Human Evolution', *Scientific American*, 1960, vol 203, no 3, pp 63–75; O J Lewis, 'Joint Remodelling in the Evolution of the Human Hand', *Journal of Anatomy*, 1977, vol 123, pp 157–201; J R Napier, 'The Prehensile Movements of the Human Hand', *Journal of Bone and Joint Surgery*, 1956, vol 38, pp 902–13; Frederick K Wood Jones, *Principles of Anatomy as seen in the Hand* (Bailliere, Tindall & Cox, 1941); Mary Marze, 'Precision Grips, Hand Morphology and Tools', *American Journal of Physical Anthropology*, 1997, vol 102, pp 91–110

Page 39 – 'Our own existence. . .'
Richard Dawkins, *The Blind Watchmaker* (Penguin, 1988)

Page 40 – But the most schematic anatomical comparison. . .
The indispensable academic text on human evolutionary anatomy is Lesley Aiello and Christopher Dean, *An Introduction to Human Evolutionary Anatomy* (Academic Press, 1990); see also Marcelline Boule and Henri Vallois, *Fossil Men: A Textbook of Human Palaeontology* (Thames & Hudson, 1957); C Owen Lovejoy, 'Evolution of Human Walking', *Scientific American*, November 1988, pp 82–9; Frederick Wood Jones, *Hallmarks of Mankind* (Bailliere, Tindall & Cox, 1948); Bruce Schechter, 'Still Standing', *New Scientist*, 14 April 2001, pp 39–42; Denis M Bramble and Daniel Lieberman, 'Endurance Running and the Evolution of Homo', *Nature*, 2004, vol 432, pp 345–51

Page 40 – So how did she come to stand upright?. . .
William Jungers, 'Lucy's Limbs: Skeletal Allometry and Locomotion in *Australopithecus afarensis*', *Nature*, 1982, vol 297, pp 676–8; Milford Wolpoff, 'Lucy's Lower Limbs: Long Enough for Lucy to be Fully Bipedal', *Nature*, 1983, vol 304, pp 59–60

Page 41 – while the *foot*, particularly the big toe. . .
W J Wang, R H Crompton, 'Analysis of the Human and Ape Foot During Bipedal Standing with Implications for the Evolution of the Foot', *Journal of Biomechanics*, 2004, vol 37, pp 1831–6; Frederick Wood Jones, *Structure and Function as Seen in the Foot* (Bailliere, Tindall & Cox, 1944)

Page 41 – There is, it would seem, an 'ideal' length. . .
Tad McGeer, 'Dynamics and Control of Bipedal Locomotion', *Journal of Theoretical Biology*, 1993, vol 163, pp 277–314; W E H Harcourt-Smith and L C Aiello, 'Fossils, Feet and the Evolution of Human Bipedal Locomotion', *Journal of Anatomy*, 2004, vol 204, pp 403–16

Page 42 – And that shortening of the arms. . .
Martin Kemp, *Leonardo da Vinci: The Marvellous Works of Nature and Man* (J M Dent & Sons, 2004)

Page 44 – The hundreds of different bones, muscles and joints. . .
D'Arcy Wentworth Thompson, *On Growth and Form* (Cambridge University Press, 1961)

Page 44 – Standing upright is, on reflection…
R H Crompton, Y Li, W Wang. 'The Mechanical Effectiveness of Erect and "Bent-hip, Bent-knee" Bipedal Walking in *Australopithacus afarensis*', *Journal of Human Evolution*, 1998, vol 35, pp 55–74; J B Saunders and Werne Inman, 'The Major Determinants in Normal and Pathological Gait', *Journal of Bone and Joint Surgery*, 1953, vol 3, pp 543–59; Neil Alexander, 'Postural Control in Older Adults', *Journal of the American Geriatric Society*, 1994, vol 42, pp 93–108; Fred Spoor and Bernard Wood, 'Implications of Early Hominid Labyrinthine Morphology for Evolution of Human Bipedal Locomotion,' *Nature*, 1994, vol 369, pp 645–7
Page 44 – These difficulties seem less acute…
This account draws heavily on John Eccles, *Evolution of the Brain* (Routledge, 1989); see also Merlin Donald, *A Mind so Rare: The Evolution of Human Consciousness* (Norton, 2001); Todd M Preuss, 'What's Human About the Human Brain?', pp 1219–34, in Michael Gazzaniga (ed), *The New Cognitive Neurosciences* (MIT Press, 2000); John Kaas, 'From Mouse to Men: The Evolution of the Large Complex Human Brain', *Journal of Biosciences*, 2005, vol 30, pp 155–65; P V Tobias, 'The Emergence of Man in Africa and Beyond', *Phil Trans R Soc Lond B*, 1981, vol 292, pp 43–56; R A Foley and P C Lee, 'Ecology and Energetics of Encephalisation in Hominid Evolution', *Phil Trans R Soc Lond B*, 1991, pp 223–32
Page 46 – The main effect of the reorientation of Lucy's pelvis…
D B Stewart, 'The Pelvis as a Passageway', *British Journal of Obstetrics and Gynaecology*, 1984, vol 91, pp 618–23; K Rosenberg and W Trevathan, 'Birth, Obstetrics and Human Evolution', *BJOG*, 2002, vol 109, pp 1199–2006; Robert Tague and C Owen Lovejoy, 'The Obstetric Pelvis of AL288–1 (Lucy)', *Journal of Human Evolution*, 1986, vol 15, pp 237–55; Harol V Jordaan, 'Newborn: Adult Brain Ratios in Hominid Evolution', *Am J Phys Anthrop*, 1976, vol 44, pp 271–8; Christopher Ruff, 'Biomechanics of the Hip and Birth in Early Homo', *Am J Phys Anthrop*, 1995, vol 98, pp 527–40
Pages 47–8 – The reports in 2006 of a family…
Uner Tan, 'A New Syndrome with Quadrupedal Gait, Primitive Speech and Severe Mental Retardation as a Live Model for Human Evolution', *International Journal of Neuroscience*, 2006, vol 116, pp 361–9
Page 48 – So, while the equivalence…
Steve Dorus et al., 'Accelerated Evolution of Nervous System Genes in the Origin of *Homo sapiens*', *Cell*, 2004, vol 119, pp 1027–40; Mario Caceres et al., 'Elevated Gene Expression Levels Distinguish Human from Non Human Primate Brains', *PNAS*, 2003, vol 100, pp 13030–5; Todd M Preuss et al., 'Human Brain Evolution: Insights from Microarrays', *Nature Reviews: Genetics*, 2004, vol 5, pp 850–60
Page 48 – 'Over the past five million years…'
Ian Tattersall (Oxford University Press, 1998), op. cit

Page 48 – The methods of the New Genetics have confirmed...
The long-standing debate over the genetic origin of modern *Homo sapiens* is well covered in two articles in *Scientific American*: Alan G Thorne and Milford Wolpoff, 'The Multi Regional Evolution of Humans', *Scientific American*, April 1992, pp 28–33; and Alan Wilson and Rebecca Cann, 'The Recent African Genesis of Humans', *Scientific American*, April 1992, pp 22–7; see also G A Clark, 'Continuity or Replacement? Putting Modern Human Origins in an Evolutionary Context', in Harold Dibble and Paul Mellars (eds), *The Middle Palaeolithic: Adaptation, Behaviour and Variability* (University of Pennsylvania, 1992). The debate was resolved in favour of a recent African origin of humans by the remarkable technical achievement of extracting DNA from fossilised Neanderthal remains. See Patricia Kahn, 'DNA from an Extinct Human', *Science*, 1997, vol 277, pp 176–8; I V Ovchinnikov et al., 'Molecular Analysis of Neanderthal DNA from the Northern Caucasus', *Nature*, 2000, vol 404, pp 490–3; David Caramelli et al., 'Evidence for a Genetic Discontinuity Between Neanderthals and 24,000-Year-Old Anatomically Modern Europeans', *PNAS*, 2003, vol 100, pp 6593–7. These issues are well covered in Chris Stringer and Robin McKie (Pimlico, 1996), op. cit.; Bryan Sykes, *Seven Daughters of Eve* (Bantam Press, 2001); Martin Jones, *The Molecule Hunt: Archaeology and the Search for Ancient DNA* (Penguin, 2001)
Page 50 – '*Homo sapiens* is not simply...'
Ian Tattersall (Oxford University Press, 1998), op. cit.
Page 50 – The precipitating factor in...
The significance of language for the emergence of human culture is well described in Philip Lieberman, *Uniquely Human: The Evolution of Speech, Thought and Selfless Behaviour* (Harvard University Press, 1991); see also Iain Davidson and William Noble, 'The Archaeology of Perception: Traces of Depiction and Language', *Current Anthropology*, 1989, vol 30, pp 125–55; Colin Renfrew and Ezra Zurbow (eds), *The Ancient Mind: Elements of Cognitive Archaeology* (Cambridge University Press, 1994); David Povinelli and Todd Preuss, 'Theory of Mind: Evolutionary History of a Cognitive Specialisation', *Trends in Neuroscience*, 1995, vol 18, pp 418–24
Page 51 – Language makes it possible...
Richard Swinburne, *The Evolution of the Soul* (Oxford University Press, 1986)
Page 51 – 'Language evolved to enable humans...'
Robin Dunbar, *The Human Story* (Faber & Faber, 2004)
Page 51 – In the 1950s the famous linguist Noam Chomsky...
Noam Chomsky, 'A Review of B F Skinner's Verbal Behaviour', *Language*, 1959, vol 35, pp 226–58; N Chomsky, *Language and the Problems of Knowledge* (MIT Press, 1988)
Page 52 – 'Within a short span of time...'
Breyne Moskowitz, 'The Acquisition of Language', *Scientific American*, November 1978, pp 82–96

Page 52 – Now, our primate cousins do not possess...
Jane Goodall, *Through a Window* (Penguin, 1990)
Page 53 – How then did the faculty of language come...
The arguments of the major protagonists in the debate over the origin of language can be found in Steven Pinker, *The Language Instinct: The New Science of Language and Mind* (Penguin, 1994), and Mark D Hauser and Noam Chomsky, 'The Faculty of Language: What is it, Who has it and How did it Evolve?', *Science,* 2002, vol 298, pp 1569–79; see also Antonio Damasio and Hanna Damasio, 'Brain and Language', *Scientific American*, September 1992, pp 63–7; Simon Fisher and Gary Marcus, 'The Eloquent Ape: Genes, Brains and Evolution of Language', *Nature Reviews: Genetics*, 2006, vol 7, pp 9–18; David Cooper, 'Broca's Arrow: Evolution Prediction, Language and the Brain', *Anatomical Record*, 2006, vol 298b, pp 9–24; Constance Holden, 'The Origin of Speech', *Science*, 2004, vol 303, pp 1316–18
Page 53 – The dispute over the evolutionary (or otherwise)...
S E Petersen, P T Fox et al., 'Positron Emission Tomography Studies of the Cortical Anatomy of Single-Word Processing', *Nature*, 1988, vol 331, pp 585–9; Michael Posner and Antonella Pavese, 'Anatomy of Word and Sentence Meaning', *Proc Nat Ac Sci*, 1998, vol 95, pp 899–905
Page 56 – Here neither of the two proposed evolutionary scenarios...
R White, 'Rethinking the Middle/Upper Palaeolithic Transition', *Current Anthropology*, 1982, vol 23, pp 169–92; P Mellars, 'Cognitive Changes and the Emergence of Modern Humans', *Cambridge Archaeological Journal*, 1991, vol 1, pp 63–76; R A Foley, 'Language Origins: The Silence of the Past', *Nature*, 1991, vol 353, pp 114–15
Page 57 – The question, rather, as the biologist Robert Wesson puts it...
Robert Wesson, *Beyond Natural Selection* (MIT Press, 1993)
Page 57 – The further subsidiary and related riddle...
The most popular 'contrary' theory attributing several distinctive human features, notably the upright stance, to an intervening 'aquatic phase' is described in Elaine Morgan, *The Scars of Evolution* (Penguin, 1990). Further critical sources include Earnest Hooton, 'Doubts and Suspicions Concerning Certain Functional Theories of Primate Evolution', *Human Biology*, 1930, vol 11, pp 223–49; John Lewis and Bernard Towers, *Naked Ape or Homo Sapiens* (Garner Press, 1969)

Chapter 3: The Limits of Science 1: The Quixotic Universe

Page 59 – The world is so full of wonder...
Michael Mayne, *This Sunrise of Wonder* (Fount, 1995); Alastair McGrath, *The Re-enchantment of Nature* (Hodder & Stoughton, 2002); Malcolm Budd, *The Aesthetic Appreciation of Nature* (Clarendon Press, 2002); E L Grant Watson, *Profitable Wonders* (Country Life Ltd, 1949)

Page 60 – 'No one can say just how many species...'
David Attenborough, *Life on Earth: A Natural History* (Collins, 1979)
Page 61 – The number of [such] admirable...
Robert Wesson (MIT Press, 1993), op. cit.
Page 62 – 'When we pause to think of the part earthworms...'
J Arthur Thomson, *The Wonder of Life* (Andrew Melrose Ltd, 1914)
Page 63 – 'There is, apart from mere intellect...'
Walt Whitman, *Specimen Days and Collected Prose* (Philadelphia, 1882)
Page 63 – 'The scientist does not study nature...'
Cited in S Chandrasekhar, *Truth and Beauty: Aesthetics and Motivation in Science* (University of Chicago Press, 1990)
Page 64 – The greatest (probably) of all scientists...
Paul Davies, *The Mind of God: Science and the Search for Ultimate Meaning* (Penguin, 1992)
Page 67 – Isaac Newton, born in 1642...
James Gleick, *Isaac Newton* (Harper Perennial, 2004)
Page 67 – 'After dinner, the weather being warm...'
William Stukeley, *Memoirs of Sir Isaac Newton's Life* (1752), quoted in John Carey (ed), *The Faber Book of Science* (Faber & Faber, 1995)
Page 69 – Newton's laws epitomise, to the highest degree...
Brian Greene, *The Elegant Universe* (Norton, 1999)
Page 70 – In the twentieth century, the conundrum...
John D Barrow and Frank J Tipler, *The Anthropic Cosmological Principle* (Oxford University Press, 1986); John Gribbin and Martin Rees, *Cosmic Coincidences, Dark Matter: Mankind and Anthropic Cosmology* (Black Swan, 1990); Paul Davies, *God and the New Physics* (Penguin, 1983); John Polkinghorne, *Beyond Science: The Wider Human Context* (Cambridge University Press, 1995)

Chapter 4: The (Evolutionary) 'Reason for Everything': Certainty

Page 72 – 'Nothing gave me so much pleasure...'
Frances Darwin, *Autobiography of Charles Darwin* (Watts & Co, 1929)
Page 72 – 'The naturalist... sees the beautiful connection...'
Editorial, *Zoological Journal*, 1824, vol 1, p 7
Page 73 – For his friend the anatomist John Hunter...
Wendy Moore, *The Knifeman* (Phantom Press, 2005)
Page 74 – The presiding genius of natural history...
G Cuvier, *Revolutions of the Surface of the Globe* (English Edition, Whittaker, Treacher & Arnot, 1829); see also William Coleman, *George Cuvier: Zoologist* (Harvard University Press, 1988)
Page 76 – The theological implications of natural history...
William Paley, *Natural Theology* (J Faulder, 1802); Peter J Bowler, 'Darwinism

and the Argument from Design: Suggestions for a Re-evaluation', *Journal of the History of Biology*, 1977, vol 10, pp 29–45

Page 78 – 'There came into my mind...'

Quoted in Alastair McGrath (Hodder & Stoughton, 2002), op. cit.

Page 78 –Perhaps, suggested the French naturalist...

J-B Lamarck, *Zoological Philosophy: An Exposition with Regard to the Natural History of Animals* (1809. Translated by Hugh Elliot, University of Chicago Press, 1984)

Page 79 – Meanwhile, Charles Darwin's enthusiasm for natural history...

Charles Darwin, *The Voyage of the Beagle* (Everyman Library, 1959); see also Alan Moorehead, *Darwin and the Beagle* (Penguin, 1971)

Page 79 – The evolutionary theory for which he is famous...

The most comprehensive biography of Darwin's life and thought is Adrian Desmond and James Moore, *Darwin* (Penguin, 1992); see also Janet Browne, *Charles Darwin: The Power of Place* (Jonathan Cape, 2006); William Coleman, *Biology in the Nineteenth Century: Problems of Form, Function and Transformation* (Cambridge University Press, 1971)

Page 81 – Most strikingly of all...

Jonathan Weiner, *The Beak of the Finch* (Jonathan Cape, 1994)

Page 83 – Fifteen months after his return...

Thomas Malthus, *An Essay on the Principle of Population* (1798, reprinted Augustus Kelley Publishers, 1986)

Page 83 – He was eventually compelled to act...

Peter Raby, *Alfred Russell Wallace: A Life* (Chatto & Windus, 2001)

Page 84 – Darwin and Newton Compared

The main sources for Darwin's evolutionary theory in this account include Mark Ridley, *Evolution* (Blackwell Scientific, 1993); Ernst Mayr, *One Long Argument* (Penguin Press, 1991); John Maynard Smith, *The Theory of Evolution* (Cambridge University Press, 1993); Michael Ruse, *The Darwin Revolution* (University of Chicago Press, 1979); Peter Bowler, *Evolution: The History of an Idea* (University of California Press, 1983); Steve Jones, *Almost Like a Whale: The Origin of the Species Updated* (Doubleday, 1999); Richard Dawkins (Penguin, 1986), op. cit.; Carl Zimmer, *Evolution* (Arrow, 2003). The most searching critique of Darwin's evolutionary theory is Michael Denton, *Evolution: A Theory in Crisis* (Adler & Adler, 1988); see also Gertrude Himmelfarb, *Darwin and the Darwinian Revolution* (Chatto & Windus, 1959); Robert Wesson (MIT Press, 1993), op. cit.; Phillip E Johnson, *Darwin on Trial* (Monarch, 1991); Brian Leith, *The Descent of Darwin* (Collins, 1982); Jonathan Wells, *Icons of Evolution* (Regnery Publishing, 2000); Francis Hitching, *The Neck of the Giraffe, or Where Darwin Went Wrong* (Pan Books, 1982); Hugh Montefiore, *The Probability of God* (SCM Press, 1985); Simon Conway Morris, *Life's Solution* (Cambridge University Press, 2003)

Page 88– A Sceptical View
Richard Owen, *Quarterly Review*, 1860, vol 108, pp 225–64 (Yale University Press, 1994); see also Nicholas Rupke, *Richard Owen, Victorian Naturalist* (Yale University Press, 1994); Peter Vorzimmer, *Charles Darwin, The Years of Controversy: The Origin of Species and its Critics, 1859–1882* (University of London Press, 1972)
Page 94 – 'Some single-cell animals...'
Richard Dawkins (Penguin, 1986), op. cit.
Page 95 – Now there is an abundance of empirical fact...
Rhona M Black, *The Elements of Palaeontology* (Cambridge University Press, 1989); Michael Denton (Adler & Adler, 1986), op. cit., chapter 8
Page 98 – The fossil record in 1859...
Martin Rudwick, *The Meaning of Fossils* (University of Chicago Press, 1972)
Page 99 – Further, the fossils of that first 'Cambrian explosion'...
Simon Conway Morris and H P Whittington, 'The Animals of the Burgess Shale', *Scientific American*, 1979, vol 24 (1), pp 101–20; Stephen J Gould, *Wonderful Life: The Burgess Shale and the Nature of History* (Hutchinson Radius, 1990)
Page 101 – 'Most people assume that...'
David Raup, 'Conflicts Between Darwin and Palaeontology', *Field Museum of Natural History Bulletin*, 1979, vol 50 (1), pp 22–9
Page 104 – The World was Ready for the Theory of Evolution
Gertrude Himmelfarb (Chatto & Windus, 1959), op. cit.
Page 104 – 'Thus we find ourselves in a singular position...'
J F Pictet, *Archives des Sciences de la Bibliothèque Universelle*, 1860, vol 3, pp 231–55, translated in D L Hull, *Darwin and his Critics* (Harvard University Press, 1973)
Page 105 – Within a decade the majority...
David Hull et al., 'Plank's Principle', *Science*, 1978, vol 2, pp 717–23
Page 105 – Indeed, the evolutionary theory could be interpreted...
Cornelius Hunter, *Darwin's God: Evolution and the Problem of Evil* (Brazos Press, 2001)
Page 105 – 'Darwin appeared under the guise of a foe...'
Aubrey Moore, 'The Christian Doctrine of God', in C G Gore (ed), *Lux Mondi* (The US Book Company, 1890)
Page 106 – The roots of the Enlightenment...
The numerous interpretations of the ideas of the Enlightenment include Norman Hampson, *The Enlightenment* (Penguin, 1968); T Z Lavine, *From Socrates to Sartre: The Philosophic Quest* (Bantam Books, 1984); Richard Tarnas, *The Passion of the Western Mind: Understanding the Ideas that have Shaped our World* (Pimlico, 1991); Gertrude Himmelfarb, *The Roads to Modernity* (Vintage, 2004)
Page 107 – 'the world to be organised...'
William Provine, 'Evolution and the Foundation of Ethics', *MBL Science*, 1988, vol 3, pp 25–9

Page 108 – 'Prior to 1859 all attempts to...'
G C Simpson, 'The Biological Nature of Man', *Science*, 1966, vol 152, pp 472–8
Page 108 – 'The sloth, instead of defecating...'
Robert Wesson (MIT Press, 1993), op. cit.
Page 109 – 'The radicalism of natural selection...'
Stephen J Gould, 'More Things in Heaven and Earth', in Hilary Rose and Steven Rose (eds), *Alas Poor Darwin: Arguments Against Evolutionary Psychology* (Jonathan Cape, 2000)

Chapter 5: The (Evolutionary) 'Reason for Everything': Doubt

Page 111 – Gregor Mendel and a (Temporary) Eclipse
John Gribbin (Allen Lane, 2002), op. cit.; Peter Bowler (University of California Press, 1983), op. cit.; Robin Marantz Henig, *A Monk and Two Peas: The Story of Gregor Mendel and the Discovery of Genetics* (Phoenix, 2000)
Page 113 – And thus, argued Professor of Genetics...
R A Fisher, *The Genetical Theory of Natural Selection* (Clarendon Press, 1930); see also Marek Kohn, *A Reason for Everything: Natural Selection and the English Imagination* (Faber & Faber, 2004)
Page 115 – 'The thoughtful student...'
James Mavor, *General Biology* (Macmillan, 1952)
Page 115 – 'The unity achieved by the...'
Margaret Morrison, *Unifying Scientific Theories* (Cambridge University Press, 2000); Anya Plutynski, 'Explanatory Unification and the Early Synthesis', *British Journal of the Philosophy of Science*, 2005, vol 56 (3), pp 595–609
Page 116 – Biologists do not, in general...
The main protagonists of the modernised version of Darwin's theory, or neo-Darwinism, include Julian Huxley, *Evolution the Modern Synthesis* (George Allen & Unwin, 1942); George Gaylord Simpson, *The Meaning of Evolution: A Study of the History of Life and its Significance for Man* (Yale University Press, 1949); Theodosius Dobzhansky, *Genetics and the Origin of Species* (Columbia University Press, 1937); Ernst Mayr, *Population, Species and Evolution* (Harvard University Press, 1970)
Page 116 – 'enable us to discern the lineaments...'
Julian Huxley in Sol Tax (ed), *Evolution after Darwin*, vol 3 (University of Chicago Press, 1960)
Page 116 – 'Possibly it is true that...'
C H Waddington, 'Theories of Evolution', in S A Barnett (ed), *A Century of Darwin* (Heinemann, 1958)
Page 117 – By 1980 the two most uncomfortable...
David Collingridge and Mark Earthy, 'Science Under Stress – Crisis in Neo-Darwinism', *Hist Phil Life Sci*, 1990, vol 12, pp 3–26; see also John Endler and Tracey McLellan, 'The Process of Evolution – Towards a Newer Synthesis', *Ann*

Rev Ecol Syst, 1988, vol 19, pp 395–421; M W Ho and P T Saunders, *Beyond Neo-Darwinism: An Introduction to the New Evolutionary Paradigm* (London Academic, 1985); Robert Wesson (MIT Press, 1993), op. cit.

Page 117 – 'For more than a century. . .'

Steven Stanley, 'Darwin Done Over', *The Sciences*, October 1981, pp 18–23

Page 118 – The main reason for palaeontologists' loss. . .

The substantial challenge to the convention and interpretation of the fossil records for Darwin's evolutionary theory is set out in Niles Eldredge, *Reinventing Darwin: The Great Evolutionary Debate* (Weidenfeld & Nicolson, 1995); see also Stephen Donovan and Christopher Paul, *The Adequacy of the Fossil Record* (John Wiley, 1998); Stephen Stanley, *Evolution: Pattern and Progress* (Johns Hopkins University Press, 1998); Stephen J Gould, 'Is a New and General Theory of Evolution Emerging?', *Palaeobiology*, 1980, vol 6, pp 119–30; Stephen J Gould, 'Darwinism and the Expansion of Evolutionary Theory', *Science*, 1982, vol 216, pp 380–7. The controversy generated by the Gould and Eldredge critique can be followed in John Maynard Smith, 'Palaeontology at the High Table', *Nature*, 1984, vol 309, pp 401–2; John Maynard Smith, 'Darwinism Stays Unpunctured', *Nature*, 1987, vol 330, p 516; S J Gould and N Eldredge, 'Punctuated Equilibria: The Tempo and Mode of Evolution Reconsidered', *Palaeobiology*, 1977, vol 3, pp 115–51; Stephen J Gould and Niles Eldredge, 'Punctuated Equilibrium Comes of Age', *Nature*, 1993, vol 366, pp 223–7

Page 120 – The conventional evolutionary scenario. . .

P D Gingerich et al., 'Origin of Whales in Epicontinental Remnant Seas', *Science*, 1983, vol 220, pp 403–6; Christian de Muizon, 'Walking with Whales', *Nature*, 2001, vol 413, pp 259–60

Page 120 – 'The central question was whether. . .'

Roger Lewin, 'Evolutionary Theory under Fire', *Science*, 1980, vol 210, pp 863–6

Page 123 – But it was clear from the moment. . .

Adam Sedgwick, 'On the Law of Development Commonly Known as von Baer's Law', *Quarterly Journal of Microscopical Science*, 1894, vol 36, pp 35–52

Page 124 – 'It does not seem to matter where. . .'

Gavin de Beer, *Homology: An Unresolved Problem* (Oxford University Press, 1971)

Page 124 – And if 'it does not matter'. . .

The most comprehensive challenge to Darwin's interpretation of the significance of homology is Jonathan Wells, *Icons of Evolution: Science or Myth*? (Regnery Publishing, 2000); see also Michael K Richardson, 'There is no Highly Conserved Embryonic Stage in Invertebrates', *Anatomy and Embryology*, 1997, vol 196, pp 91–106; Gregory Wray and Ehab Abouheif, 'When is Homology not Homology?', *Current Opinion in Genetics and Development*, 1998, vol 8, pp 675–80

Page 125 – But behind the façade of scientific. . .

G Ledyard Stebbins and Francisco Ayala, 'Is a New Evolutionary Synthesis Necessary?', *Science*, 1981, vol 213, pp 967–72; Stephen J Gould, 'Is a New and General Theory of Evolution Emerging?', *Palaeobiology*, 1980, vol 6, pp 119–30.

This protracted controversy is well covered in Andrew Brown, *The Darwin Wars* (Simon & Schuster, 1999); Kim Sterelny, *Dawkins versus Gould: Survival of the Fittest* (Icon, 2001)

Chapter 6: The Limits of Science 2: The Impenetrable Helix

Page 126 – 'It seems as though God has . . .'
Nicolas Malebranche, *The Search after Truth* (Ohio State University Press, 1980)
Page 127 – 'We would see endless corridors. . .'
Michael Denton (Adler & Adler, 1986), op. cit.
Page 128 – The humble fly is. . .
Walter M Elsasser, *Reflections on a Theory of Organisms* (Johns Hopkins University Press, 1988)
Page 129 – The second instalment began in 1953. . .
The several historical accounts of the discovery of the Double Helix and its aftermath include Walter Bodmer and Robin McKie, *The Book of Man* (Little, Brown, 1994); Michael Morange, *A History of Molecular Biology* (Harvard University Press, 1998); Enrico Coen, *The Art of Genes* (Oxford University Press, 1999); Lily E Kay (Stanford University Press, 2000), op. cit.; Henry Gee, *Jacob's Ladder: The History of the Human Genome* (Fourth Estate, 2004)
Page 129 – 'I have just with some difficulty. . .'
Christopher Wills, *Exons, Introns and Talking Genes: The Science Behind the Human Genome Project* (Oxford University Press, 1992)
Page 130 – 'We imagine that, prior to duplication. . .'
J D Watson and F C Crick, 'Molecular Structure of Nucleic Acids: A Structure for Deoxyribose Nucleic Acid', *Nature*, 1953, vol 171, p 737
Page 131 – We turn now to the crucial question. . .
An accessible, fully illustrated description of the science behind the New Genetics can be found in James D Watson, *DNA: The Secret of Life* (William Heinemann, 2003)
Page 131 – We will focus on just one. . .
The intricacies of proteins and their structure are well described in Charles Tanford and Jacqueline Reynolds, *Nature's Robots: A History of Proteins* (Oxford University Press, 2001); David Goodsell, *Our Molecular Nature: The Body's Motors, Machines and Messages* (Copernicus, 1996)
Page 135 – These dual powers confer on. . .
Evelyn Fox Keller, *The Century of the Gene* (Harvard University Press, 2000); Dorothy Nelkin and M Susan Lindee, *The DNA Mystique: The Gene as a Cultural Icon* (W H Freeman & Co, 1995)
Page 135 – 'On the one hand [there is]. . .'
M A Edey and D C Johanson, *Blueprints: Solving the Mystery of Evolution* (Oxford University Press, 1990)

Page 135 – The first of those innovations...
James Le Fanu (Little, Brown, 1999), op. cit.
Page 137 – The red blood cells in those...
R M Hardisty and D J Wetherall (eds), *Blood and its Disorders* (Blackwell Scientific, 1982); D J Wetherall, *The New Genetics in Clinical Practice* (Nuffield Provincial Hospitals Trust, 1982)
Page 138 – Many genetic illnesses proved to...
L C Tsiu, 'The Spectrum of Cystic Fibrosis Mutations', *Trends in Genetics*, 1992, vol 8, pp 392–8
Page 138 – Further, it emerged, it was...
David Papermaster, 'Necessary but Insufficient', *Nature Medicine*, 1995, vol 1, pp 874–5
Page 138 – These complexities, inexplicable in the...
Ulrich Wolf, 'Identical Mutations and Phenotypic Variation', *Human Genetics*, 1997, vol 100, pp 305–21
Page 139 – The puzzle proliferated further when...
Jan A Witkowskie, 'Manipulating DNA: From Cloning to Knockouts', in Margery Ord and Lloyd Stocken (eds), *Quantum Leaps in Biochemistry* (Jai Press Inc, 1996)
Page 139 – 'In some cases a "knockout" may...'
Michel Morange, *The Misunderstood Gene* (Harvard University Press, 2001)
Page 140 – 'The heart of the problem...'
P G H Gell, 'Destiny of the Genes', in R Duncan and M Weston-Smith (eds), *The Encyclopedia of Medical Ignorance* (Pergamon, 1984)
Page 140 – And sure enough, in the late...
Walter Gehring, *Master Control Genes in the Development of Evolution: The Homeobox Story* (Yale University Press, 1999)
Page 140 – But when Gehring and his colleagues...
Henry Gee (Fourth Estate, 2004), op. cit., contains an excellent discussion of the significance of the Hox genes; the first papers reporting their interchangeability between species include Alexander Awgulewitsch and Donna Jacobs, 'Deformed Autoregulatory Elements from Drosophila function in a Conserved Manner in Transgenic Mice', *Nature*, 1992, vol 358, pp 341–5; Jarema Malicki, 'A Human HOX 4B Regulatory Element Provides Head-Specific Expression in Drosophila Embryos', *Nature*, 1992, vol 358, pp 345–7
Page 141 – It is possible to get a feel for...
Sean B Carroll, *The Making of the Fittest* (Norton, 2006)
Page 143 – The implications of this conundrum are...
Sean B Carroll, *Endless Forms Most Beautiful* (Weidenfeld & Nicolson, 2005). The specific references for experiments mentioned in this passage include: R D Fernald, 'Evolution of Eyes', *Current Opinion*, in *Neurobiology*, 2000, vol 10, pp 444–50; R Quiring et al., 'Homology of the Eyeless Gene of Drosophila to the Small Eye Gene in Mice and Aniridia in Humans', *Science*, 1994, vol 265, pp 785–9;

G Halder, 'Induction of Ectopic Eyes by Targeted Expression of the Eyeless Gene in Drosophila', *Science*, 1995, vol 267, pp 178–92; R L Chow et al., 'Pax 6 Induces Ectopic Eyes', *Invertebrate Development*, 1999, vol 126, pp 4213–22; G Panganiban et al., 'The Origins and Evolution of Animal Appendages', *Proceedings of the National Academy of Science*, 1997, vol 94, pp 5162–6; R Bodmer and T V Venkatesh, 'Heart Development in Drosophila and Vertebrates: Conservation of Molecular Mechanisms', *Developmental Genetics*, 1998, vol 22, pp 181–6

Page 146 – The radical implications of that...

The limits of genetic explanation and inference of hidden biological forces imposing the order of form are explored in E S Russell, *Form and Function: A Contribution to the History of Animal Morphology* (John Murray, 1916); Rupert Sheldrake, *The Presence of the Past* (Park Street Press, 1988); Gerry Webster and Brian Goodwin, *Form and Transformation: Generative and Relational Principles in Biology* (Cambridge University Press, 1996); Stanley Shostack, *The Legacy of Molecular Biology* (Macmillan, 1998); Lenny Moss, *What Genes Can't Do* (MIT Press, 2003); Susan Oyama, *Evolution's Eye: A Systems View of the Biology–Culture Divide* (Duke University Press, 2000)

Chapter 7: The Fall of Man: A Tragedy in Two Acts

Page 148 – 'This moving and sublime spectacle of nature...'

Denis Diderot, *Encyclopédie ou Dictionnaire Raisonné des Sciences, des Arts, des Metiers* (1752–65)

Page 149 – 'When I endeavour to examine my...'

Adam Smith, *The Theory of Modern Sentiments*, ed D D Raphael and A I Macfie (Oxford University Press, 1976; reprint of sixth edition, 1790)

Page 150 – 'That little boy with the shock of...'

John Polkinghorne, *Beyond Science: The Wider Human Context* (Cambridge University Press, 1995)

Page 150 – That self, the 'inner person'...

Richard Swinburne, *The Evolution of the Soul* (Clarendon Press, 1986)

Page 150 – This conundrum of the dual character...

Kenan Malik, *Man, Beast and Zombie: What Science Can and Cannot Tell us About Human Nature* (Weidenfeld & Nicolson, 2000)

Page 152 – The most persuasive argument for man's evolutionary...

Charles Darwin, *The Descent of Man* (John Murray, 1871)

Page 155 – 'The astonishment which I felt on...'

Charles Darwin, *The Voyage of the Beagle* (Everyman Library, J M Dent & Sons, 1959)

Page 155 – 'There exists in man some close...'

J Barnard Davis, 'Contributions Towards Determining the Weight of the Brain in the Different Races of Man', *Royal Society of London Proceedings*, 16 (1867–68), pp 236–41

Page 156 – 'Darwin had no adequate concept of Man'

John Greene, *Debating Darwin* (Regina Books, 1999); Gertrude Himmelfarb (Chatto & Windus, 1959), op. cit.; Howard Gruber, *Darwin on Man* (Wildwood House, 1974); Tim Lewens, *Darwin* (Routledge, 2007)

Page 156 – 'After a long and intimate intercourse...'

Franz Boas, *Anthropology and Modern Life* (David Publications, 1988)

Page 156 – 'Life *was* a struggle: every businessman...'

C E Raven, *Natural Religion and Christian Theology* (Cambridge University Press, 1953)

Page 159 – There were others, however, who...

Richard Weikart, *From Darwin to Hitler: Evolutionary Ethics, Eugenics and Racism in Germany* (Palgrave Macmillan, 2004); Elof Axel Carlson, *The Unfit: The History of a Bad Idea* (Cold Spring Harbor Laboratory Press, 2001); Richard Lynn, *Eugenics: A Reassessment* (Praeger, 2001); Daniel Jo Kevles, *In the Name of Eugenics* (Penguin, 1985)

Page 159 – 'If a twentieth part of the cost and pains...'

Francis Galton, 'Hereditary Character and Talent', *MacMillan's Magazine*, 1864, vol 11, pp 157–66

Page 160 – 'In Virginia, the state sterilisation authorities...'

Cited in Daniel Kevles (Penguin, 1985), op. cit.

Page 160 – 'What profit does humanity derive from...'

Ernst Haeckel, *Die Lebenswunder* [The Wonders of Life] (Alfred Kroner, 1904)

Page 160 – Paralleling the new science of eugenics...

John S Haller, *Outcasts from Evolution: Scientific Attitudes of Racial Inferiority 1859–1900* (Southern Illinois University Press, 1971)

Page 162 – 'The creed of the Allmacht...'

Vernon Kellog, *Headquarters Nights: A Record of Conversations and Experiences at the Headquarters of the German Army in France and Belgium* (Atlantic Monthly Press, 1917)

Page 163 – 'In such a system, human life...'

Roger Scruton, *An Intelligent Person's Guide to Philosophy* (Duckworth, 1997)

Page 163 – They 'come with the territory'...

Richard Weikart (Palgrave Macmillan, 2004), op. cit.

Page 164 – And many, many more, not least...

Matt Ridley, *Francis Crick* (Atlas Books, 2006)

Page 165 – 'They [the genes] swarm in huge colonies...'

Richard Dawkins, *The Selfish Gene* (Oxford University Press, 1976)

Page 165– This most unusual (and provocative) take on...

Edward O Wilson, *Sociobiology: The New Synthesis* (Harvard University Press, 1975); E O Wilson, *On Human Nature* (Penguin, 1995; first published 1978)

Page 166 – 'Altruism ought to be non-existent...'
David Stove, *Darwinian Fairy Tales* (Avebury, 1995)
Page 167 – While an undergraduate at Cambridge University...
Marek Kohn (Faber & Faber, 2004), op. cit.
Page 167 – Hamilton found a way to express this idea...
W D Hamilton, 'The Genetical Evolution of Social Behaviour', *Journal of Theoretical Biology*, 1964, vol 7, pp 1–52
Page 169 – 'Assume that the chance of the man drowning...'
Robert Trivers, 'The Evolution of Reciprocal Altruism', *Quarterly Review of Biology*, 1971, vol 46, pp 35–57
Page 170 – 'The economy of nature is competitive from...'
M T Ghiselin, *The Economy of Nature and the Evolution of Sex* (University of California Press, 1974)
Page 171 – Wilson claims, in short, that we are deluded...
The most trenchant rebuttal of Wilson's claims and their implications is Kenan Malik (Weidenfeld & Nicolson, 2000), op. cit.; see also Steven Rose and R C Lewontin, *Not in our Genes: Biology, Ideology and Human Nature* (Penguin, 1984); Mary Midgley, *Beast and Man: The Roots of Human Nature* (Cornell University Press, 1978); Mikael Stenmark, *Scientism, Science, Ethics and Religion* (Ashgate, 2001); Andrew Brown, *The Darwin Wars: How Stupid Genes Became Selfish Gods* (Simon & Schuster, 1999)
Page 171 – Wilson's sociobiology would in turn spawn...
The foundational text of evolutionary psychology is Jerome Barkow, Leda Cosmides and John Tooby (eds), *The Adapted Mind: Evolutionary Psychology and the Generation of Culture* (Oxford University Press, 1992); see also Alan Clamp, *Evolutionary Psychology* (Hodder & Stoughton, 2000); Steven Pinker, *How the Mind Works* (Allen Lane, 1997); Henry Plotkin, *Evolution in Mind* (Allen Lane, 1997); Robin Dunbar, *The Human Story: A New History of Mankind's Evolution* (Faber & Faber, 2004); David Buss, *The Evolution of Desire: Strategies of Human Mating* (Basic Books, 1994); Robert Wright, *The Moral Animal: The New Science of Evolutionary Psychology* (Vintage, 1994); Matt Ridley, *The Origins of Virtue* (Viking, 1996); Helena Cronin, *The Ant and the Peacock* (Cambridge University Press, 1991). The several critiques of evolutionary psychology include: Susan McKinnon, *Neo-Liberal Genetics: The Myths and Moral Tales of Evolutionary Psychology* (Prickly Paradigm Press, 2005); see also Hilary Rose and Steven Rose (Jonathan Cape, 2000), op. cit.; Jerry Fodor, 'The Trouble with Psychological Darwinism', *London Review of Books*, 22 January 1998, pp 11–13
Page 172 – Women, on the other hand...
John Cartwright, *Evolution and Human Behavior* (MIT Press, 2000)
Page 173 – Some men, of course, are not up to...
Randy Thornhill and Craig Palmer, *The Natural History of Rape: Biological Bases of Sexual Coercion* (MIT Press, 2000)

Page 173 – 'a shocking attempt to ensnare us all. . .'
J C King, 'Sociobiology: Are Values and Ethics Determined by the Gene?', in Ashley Montagu (ed), *Sociobiology Examined* (Oxford University Press, 1980)
Page 173 – 'No one has ever been able to relate any aspect. . .'
Richard Lewontin, 'Sociobiology: Another Biological Determinism', *International Journal of Health Services*, 1980, vol 10, p 347; see also Charles Mann, 'Behavioral Genetics in Transition', *Science*, 1994, vol 264, pp 1686–9

Chapter 8: The Limits of Science 3: The Unfathomable Brain

Page 177 – The significance of human individuality. . .
Gordon Allport, *Becoming: Basic Considerations for a Psychology of Personality* (Yale University Press, 1978)
Page 179 – 'You, your joys and your sorrows. . .'
Francis Crick, *The Astonishing Hypothesis: The Scientific Search for the Soul* (Simon & Schuster, 1994)
Page 180 – The external map of the brain. . .
The two best short introductions to the brain are Colin Blakemore, *Mechanics of the Mind* (Cambridge University Press, 1977), and Susan Greenfield, *The Human Brain: A Guided Tour* (Weidenfeld & Nicolson, 1997)
Page 180 – The casualties of the trenches. . .
Mitchell Glickstein, 'The Discovery of the Visual Cortex', *Scientific American*, September 1988, pp 84–91
Page 180 – From the 1930s onwards the. . .
W Penfield and T Rasmussen, *The Cerebral Cortex of Man: A Clinical Study of Localisation of Function* (Macmillan, 1957)
Page 182 – The parcelling up of the real estate. . .
The implications of the localisation of function in the discrete parts of the brain are explored in Susan Leigh-Star, *Regions of the Mind: Brain Research and the Quest for Scientific Certainty* (Stanford University Press, 1989); see also Martha J Farah, 'Neuropsychological Inference with an Interactive Brain: A Critique of the "Locality" Assumption', *Behavioral and Brain Sciences*, 1994, vol 17, pp 43–104; Darryl Bruce, 'Fifty Years Since Lashley's *In Search of the Engram*: Refutations and Conjectures', *Journal of the History of the Neurosciences*, 2001, vol 10, pp 308–18
Page 182 – These integrative functions of the frontal lobes. . .
Malcolm MacMillan, *An Odd Kind of Fame: Stories of Phineas Gage* (MIT Press, 2000)
Page 183 – Both computers and brains are. . .
For a very useful account of neuroscience and the brain/computer analogy see John McCrone (Faber & Faber, 1999), op. cit.; and see also Kenan Malik (Weidenfeld & Nicolson, 2000), op. cit.; Howard Gardner, *The Mind's New*

Science: A History of the Cognitive Revolution (Basic Books, 1987); Michael Gazzaniga (ed) (MIT Press, 2001), op. cit.; Stephen M Kosslyn and Olivier Koenig, *Wet Mind: The New Cognitive Neuroscience* (The Free Press, 1992); John Horgan, *The Undiscovered Mind: How the Brain Defies Explanation* (Weidenfeld & Nicolson, 1999); Merlin Donald (Norton, 2001), op. cit.; Adam Zeman, *Consciousness: A User's Guide* (Yale University Press, 2002)

Page 186 – 'A computer is a machine for processing...'
Raymond Tallis, *The Explicit Animal: A Defence of Human Consciousness* (Macmillan, 1991)

Page 187 – 'We don't just have the power of...'
Charles Jonscher, *Wired Life: Who are We in the Digital Age?* (Bantam, 1990)

Page 187 – The notion of human individuality...
Matt Ridley, *Nature via Nurture: Genes, Experience and What Makes us Human* (Fourth Estate, 2003)

Page 187 – 'There is no escape that nature...'
Francis Galton, *Hereditary Genius: An Inquiry into its Laws and Consequences* (Macmillan, 1869)

Page 188 – A century later...
T J Bouchard et al., 'Sources of Human Psychological Differences: The Minnesota Study of Twins Reared Apart', *Science*, 1990, vol 250, pp 223–8

Page 188 – 'Each was wearing a beige dress...'
Lawrence Wright, *Twins* (Phoenix, 1998)

Page 189 – 'For a child to learn the meaning of words...'
Lawrence Hirschfeld and S A Gelman (eds), *Mapping the Mind* (Cambridge University Press, 1994)

Page 190 – 'a baby crawling from New York to Seattle...'
Jeffrey Schwartz, *The Mind and the Brain* (Regan Books, 2002)

Page 190 – Now we turn to the second element...
This discussion of neuroplasticity draws heavily on Jeffrey Schwartz (Regan Books, 2002), op. cit., and Matt Ridley (Fourth Estate, 2003), op. cit.; see also Patrick Bateson and Paul Martin, *Design for Life: How Behaviour Develops* (Jonathan Cape, 1999); Michael Rutter, *Genes and Behaviour: Nature–Nurture Interplay Explained* (Blackwell, 2006)

Page 190 – The American anthropologist...
Ashley Montagu, *Man in Process* (World Publishing Co, 1961)

Page 190 – Similarly, when in the post-war years...
W J Kim, 'International Adoption: A Case Review of Korean Children', *Child Psychiatry and Human Development*, 1995, vol 25, pp 141–54

Page 191 – 'We wanted to have a rough idea of...'
David Hubel and Torsten Wiesel, *Brain and Visual Perception* (Oxford University Press, 2005)

Page 192 – Thus the brain scan of a previously...
Leonard Yuen, *BMJ*, 2003, vol 327, p 998

Page 194 – 'We are able to identify objects...'
Raymond Tallis (Macmillan, 1991), op. cit.
Page 194 – There has to be, it must be presumed...
John Horgan, 'Will Anyone Ever Decode the Human Brain?', *Discover*, 2004, vol 25; see also Graeme Mitchinson, 'The Enigma of the Cortical Code', *Trends in Neuroscience*, 1990, vol 13, pp 41–3; J J Eggermont, 'Is there a Neural Code?', *Neurosci Biobehav Rev*, 1998, vol 22, pp 355–700
Page 195 – 'Standing by a pond in London Zoo...'
John McCrone (Faber & Faber, 1999), op. cit.
Page 195 – But there is no mistaking the...
S E Petersen, P T Fox, M I Posner, M E Raichle, 'Positron Emission Tomography Studies of the Cortical Activity of Single-Word Processing', *Nature*, 1988, vol 331, pp 585–9
Page 196 – 'Astronomers were building huge telescopes...'
John McCrone (Faber & Faber, 1999), op. cit.
Page 198 – Ten years on, in 1998, when...
'Papers from a National Academy of Sciences Colloquium on Neuroimaging of Human Brain Function', *Proc Nat Ac Sci*, 1998, vol 95, pp 763–1348
Page 198 – It would be a hopeless task to attempt...
The most comprehensive survey of the findings of the Decade of the Brain is Robert Cabeza and Alan Kingstone (eds), *Handbook of Functional Neuroimaging of Cognition* (MIT Press, 2001); see also Pert Rowland, *Brain Activation* (Wyley-Liss, 1993); Peter Jezzard, Paul Matthews, Stephen Smith, *Functional MRI: An Introduction to Methods* (Oxford University Press, 2002); William Uttal (MIT Press, 2001), op. cit.; M I Posner and M E Raichle *Images of the Mind* (Scientific American Library, 1994); Rita Carter, *Mapping the Mind* (Phoenix, 1998)
Page 199 – Further, and this is yet more difficult...
W D Wright, *The Rays are not Coloured: Essays on the Science of Vision and Colour* (Adam Hilger, 1967); see also Rupert Sheldrake, *The Sense of Being Stared at and Other Aspects of the Extended Mind* (Hutchinson, 2003)
Page 200 – 'One stormy night when my nephew...'
Rachel Carson, *The Sense of Wonder* (Harper & Row, 1965)
Page 200 – 'As surely as the old system of beliefs...'
Semir Zeki (Blackwell Scientific, 1993), op. cit.
Page 201 – 'Here begins the brain's 'quiet conversation...'
Melvyn Goodale, 'Perception and Action in the Human Visual System', in Michael Gazzaniga (ed) (MIT Press, 2001), op. cit. A non-technical account of the neuroscience of vision can be found in Thomas B Czerner, *What Makes you Tick: The Brain in Plain English* (John Wiley, 2001)
Page 202 – But the reverse is the case...
T J Fellerman and D C Van Essen, 'Distributed Hierarchical Processing in the Primate Cerebral Cortex', *Cerebral Cortex*, 1991, vol 1, pp 1–47

Page 202 – Professor Zeki speculated that two...
Semir Zeki (Blackwell Scientific, 1993), op. cit.; see also Semir Zeki, 'The Visual Image of Mind and Brain', *Scientific American*, September 1992, pp 43–50; Semir Zeki, 'A Direct Demonstration of Functional Specialisation in Human Visual Cortex', *Journal of Neuroscience*, 1991, vol 11, pp 641–9
Page 202 – The precise function of many of those...
A R McIntosh et al., 'Network Analysis of Cortical Visual Pathways Mapped with PET', *Journal of Neuroscience*,1994, vol 14, pp 655–6
Page 205 – They exemplify too the progressively mounting...
Zenon Pylyshyn, 'Is Vision Continuous with Cognition? The Case for Cognitive Impenetrability of Visual Perception', *Behavioral and Brain Sciences*, 1999, vol 22, pp 341–423; see also Daniel Pollen, 'On the Neural Correlates of Visual Perception', *Cerebral Cortex*, 1999, vol 9, pp 4–19
Page 206 – 'This abiding tendency for attributes such as...'
David Hubel (Scientific American Library, 1988), op. cit.
Page 208 – 'Henry works in a state rehabilitation centre...'
Colin Blakemore (Cambridge University Press, 1977), op. cit.
Page 209 – But it took just a few moments of rehearsal...
Marcus Raichle et al., 'Practice Related Changes in Human Brain Function Anatomy During Non-Motor Learning', *Cerebral Cortex*, 1994, vol 4, pp 8–26
Page 210 – The next and obvious form of investigation...
Alex Martin, 'The Functional Neuroimaging of Semantic Memory'; John Gabrieli, 'Functional Imaging of Episodic Memory'; Mark D'Esposito, 'Functional Neuroimaging of Working Memory', in Robert Cabeza and Alan Kingstone (eds) (MIT Press, 2001), op. cit.
Page 211 – 'Dr P was a musician of distinction...'
Oliver Sacks, *The Man Who Mistook his Wife for a Hat* (Picador, 1985)
Page 211 – A London taxi driver, by contrast...
Eleanor A Maguire et al., 'Recalling Routes Around London', *Journal of Neuroscience*, 1997, vol 17, pp 7103–10
Page 211 – Both those very distinct forms of...
Eleanor A Maguire, Christopher Frith, 'Ageing Affects the Engagement of the Hippocanthus during Autobiographical Memory Retrieval', *Brain*, 2003, vol 126, pp 1511–23
Page 212 – 'It is worthwhile to consider why...'
Randy Buckner, 'Neuroimaging of Memory', in Michael S Gazzaniga (ed) (MIT Press, 2001), op. cit.
Page 213 – Subsequently, neuroscientists in Toronto...
R Cabeza et al., 'Age Related Differences in Effective Neural Connectivity During Encoding and Recall', *NeuroReport*, 1997, vol 8, pp 3479–83
Page 214 – these changes in the brain's connections...
Lars Nyberg, 'Common Prefrontal Activations During Working Memory, Episodic Memory and Semantic Memory', *Neuropsychologia*, 2003, vol 41,

pp 371–7; see also Charan Ranganath et al., 'Prefrontal Activity Associated with Working Memory and Episodic Long Term Memory', *Neuropsychologia*, 2003, vol 41, pp 378–89

Page 214 – The simplest way of conceiving of a memory...

Eric Kandel and Robert D Hawkins, 'The Biological Basis of Learning and Individuality', *Scientific American*, September 1992, pp 53–60; see also Eric Kandel and Christopher Pittenger, 'The Past, the Future and the Biology of Memory Storage', *Phil Trans R Soc Lond B*, 1999, vol 354, pp 2027–52; Larry Squire and Eric Kandel, *Memory from Mind to Molecules* (Scientific American Library, 1999); Eric Kandel, *In Search of Memory* (Norton, 2006); Natalie Tronson and Jane Taylor, 'Molecular Mechanisms of Memory Reconsolidation', *Nature Review: Neuroscience*, 2007, vol 8, pp 262–75

Page 218 – Free will poses a 'higher order' difficulty...

The neuroscience of free will is lucidly explained in Jeffrey Schwartz (Regan Books, 2002), op. cit.; see also Graham Dunstan Martin, *Does it Matter? The Unsustainable World of the Materialists* (Floris Books, 2005); Richard Swinburne, *The Evolution of Soul* (Clarendon Press, 1986); Max Velmans, 'How Could Conscious Experiences Affect Brains', *Journal of Consciousness Studies*, 2002, vol 9, pp 3–29

Page 219 – 'Free will as it is traditionally conceived...'

William Provine, 'Evolution and the Foundation of Ethics', *MBL Science*, 1988, vol 3, pp 25–9

Page 219 – 'No object can *catch* our attention except...'

William James, *The Principles of Psychology* (Harvard University Press, 1983)

Page 220 – There are few simpler brain-imaging tasks...

John Reynolds, 'How are Features of Objects Integrated into Perpetual Wholes that are Selected by Attention?', in J Leo van Hemmen and Terrence J Sejnowski, *Twenty-Three Problems in Systems Neuroscience* (Oxford University Press, 2006)

Page 220 – Thus the self 'pays attention' by...

Ian H Robertson, *Mind Sculpture: Your Brain's Untapped Potential* (Phantom Press, 1999)

Page 220 – 'Anna was twenty-four, a graduate student...'

Jeffrey Schwartz (Regan Books, 2002), op. cit.

Page 221 – And sure enough, Professor Schwartz...

Jeffrey Schwartz et al., 'Systematic Changes in Cerebral Glucose Metabolic Rate After Successful Behaviour Modification Treatment of Obsessive Compulsive Disorder', *Arch Gen Psychiatry*, 1996, vol 53, pp 109–13

Page 222 – Professor Schwartz's discovery inspired...

Vincent Paquette et al., 'Change the Mind and you Change the Brain: Effects of Cognitive-Behavioural Therapy on the Neural Correlates of Spider Phobia', *NeuroImage*, 2003, vol 18, pp 401–9

Page 222 – 'Collectively the findings of these studies...'

Mario Beauregard et al., 'Mind Really Does Matter: Evidence from

Neuroimaging Studies of Emotional Self Regulation, Psychotherapy and Placebo Effect', *Progress in Neurobiology* 2007, vol 81, pp 482–761
Page 222 – For the pioneer of PET scanning...
M I Posner and M E Raichle (Scientific American Library, 1994), op. cit.
Page 223 – His collegue Marcus Raichle concurs...
Marcus Raichle, 'Behind the Scenes of Functional Brain Imaging: A Historical and Physiological Perspective', *Proc Nat Ac* Sci, 1998, vol 95, pp 765–72
Page 223 – Finally, the 'Big Science' of neuroscience...
William Uttal (MIT Press, 2001), op. cit.; see also Karl Friston, 'Beyond Phrenology: What Can Neuroimaging Tell us About Distributed Circuitry?', *Ann Rev Neurosci*, 2002, vol 25, pp 221–50
Page 224 – 'Conscious intelligence is a wholly natural...'
Paul Churchland, *Matter and Consciousness* (MIT Press, 1988)
Page 224 – 'Conscious human minds are more-or-less...'
Daniel Dennett, *Consciousness Explained* (Viking, 1991)
Page 225 – '[The mind] is simply a higher-level or...'
John Searle, *The Rediscovery of the Mind* (MIT Press, 1992)
Page 225 – 'The bond between mind and brain...'
Colin McGinn, *Mysterious Flame: Conscious Minds in a Material World* (Basic Books, 1999)
Page 226 – 'The seemingly limitless and enduring...'
Robert Doty, 'The Five Mysteries of the Mind and their Consequences', *Neuropsychologia*, 1998, vol 36, pp 1069–76
Page 227 – Next, those five cardinal mysteries...
The history of the human soul is recounted in Rosalie Osmond, *Imagining the Soul: A History* (Sutton Publishing, 2003); see also John Foster, *The Immaterial Self: A Defence of the Cartesian Dualist Conception of the Mind* (Routledge, 1991); W Jones, 'Brain, Mind and Spirit: Why I am not Afraid of Dualism', in Kelly Bulkeley (ed), *Soul, Psyche and Brain: New Directions in the Study of Religion and Brain Mind Science* (Palgrave Macmillan, 2005); Mario Beaureguard, *The Spiritual Brain* (HarperOne, 2007); Richard Swinburne (Clarendon Press, 1986), op. cit.; Graham Dunstan Martin (Floris Books, 2005), op. cit.

Chapter 9: The Silence

Page 231 – The journal *Science*, commenting on ...
Eric H Davidson, 'The Sea Urchin Genome: Where Will it Lead Us?', *Science*, 2006, vol 314, pp 93–4
Page 231 – 'The human brain is a machine...'
Colin Blakemore, *The Mind Machine* (BBC, 1988)
Page 232 – 'Darwin's theses of common descent...'
Ernst Mayr, *Scientific American*, July 2000, pp 67–71

Page 232 – 'It is not that the methods and institutions...'
R C Lewontin, 'Billions and Billions of Demons', *New York Review of Books*, 9 January 1997
Page 233 – It is much better, as the writer and biologist...
Stephen J Gould, *Rocks of Ages: Science and Religion in the Fullness of Life* (Jonathan Cape, 2001)
Page 234 – 'Although many details remain to be worked out...'
George Gaylord Simpson, *The Meaning of Evolution* (Yale University Press, 1967)
Page 234 – 'There is darkness without and when I die...'
Bertrand Russell, *Mysticism and Logic* (Longmans Green, 1918)
Page 235 – We are so familiar with, and committed to...
Antony O'Hear, *After Progress: Finding the Old Way Forward* (Bloomsbury, 1999)
Page 235 – 'Science suddenly stood forth as mankind's...'
Richard Tarnas, *The Passion of the Western Mind* (Pimlico, 1991)
Page 235 – The findings of the scientific revolution...
Gertrud Himmelfarb (Vintage, 2004), op. cit.; see also Thomas Hankins, *Science and the Enlightenment* (Cambridge University Press, 1985)
Page 236 – 'We are intelligent beings...'
Voltaire, *The Philosophical Dictionary*, selected and translated by H I Woolf (Knopf, 1924)
Page 237 – The origins of the microscope stretch back to...
Brian Bracegirdle, 'The Microscopical Tradition', in W F Bynum and Roy Porter, *Companion Encyclopedia of the History of Medicine* (Routledge, 1993); see also Catherine Wilson, *The Invisible World: Early Modern Philosophy and the Invention of the Microscope* (Princeton University Press, 1995)
Page 237 – This previously hidden world...
E Travnikova, 'Jan Evangelista Purkinje 1787–1869', *Physiologia Bohemoslavaca*, 1987, vol 36, pp 181–9
Page 237 – Along the way, in 1839...
Henry Harris, *The Birth of the Cell* (Yale University Press, 1999)
Page 239 – The enthusiastic microscopists then turned...
William Coleman, *Biology in the Nineteenth Century: Problems of Form, Function and Transformation* (Cambridge University Press, 1971)
Page 239 – Paralleling this remorseless unravelling...
John Gribbin (Allen Lane, 2002), op. cit.; see also William H Brock, *Fontana History of Chemistry* (Fontana Press, 1992)
Page 242 – 'The vital [animating] force does not exist...'
Émil du Bois-Reymond, *I Confini della Conoscenza della Natura* (Feltrinelli, 1973); cited in Giovanni Federspil, 'The Nature of Life in the History of Medical and Philosophical Thinking', *Am J Nephrol*, 1994, vol 14, pp 337–43

Page 242 – 'These attempts at the division...'
Cited in Fritjof Capra, *The Web of Life* (Flamingo, 1997)
Page 244 – 'The spiritual had sunk the human race in...'
Graham Dunstan Martin, *Does it Matter?* (Floris Books, 2005). The contribution of science to the secularisation of Western thought is examined in Owen Chadwick, *The Secularisation of the European Mind in the Nineteenth Century* (Cambridge University Press, 1975); see also Frederick Gregory, 'The Impact of Darwinian Evolution on Protestant Theology in the Nineteenth Century', in David Lindberg and Ronald Numbers (eds), *God and Nature: Historical Essays on the Encounter Between Christianity and Science* (University of California Press, 1986); John Hedley Brooke, *Science and Religion: Some Historical Perspectives* (Cambridge University Press, 1991); A N Wilson, *God's Funeral* (John Murray, 1999)
Page 245 – 'the humble and submissive slave of a...'
Ludwig Büchner, *Force and Matter* (1855), cited in Owen Chadwick (Cambridge University Press, 1975), op. cit.
Page 246 – 'I was present at a meeting of one...'
Robert Augros and George Stanciu, *The New Biology: Discovering the Wisdom in Nature* (New Science Library, 1988)
Page 246 – 'Nobody has the slightest idea how anything...'
Jerry Fodor, 'The Big Idea', *Times Literary Supplement*, 3 July 1992
Page 247 – The perception that science's glory days...
John Horgan, *The End of Science* (Addison Wesley, 1996); see also Bentley Glass, 'Science: Endless Horizons or Golden Age?', *Science*, 8 January 1971, pp 23–9
Page 247 – So too medicine, where the tidal wave...
James Le Fanu (Little, Brown, 1999), op. cit.
Page 248 – But in general the fortunes of the biotechnology industry...
Editorial, 'Biotech's Uncertain Future', *Lancet*, 1996, vol 437, p 1497; see also Andrew Pollack, 'It's Alive! Meet One of Biotech's Zombies', *New York Times*, 11 February 2007; R C Lewontin, 'The Dream of the Human Genome', *New York Review of Books*, 18 May 1992
Page 249 – The overwhelming impression is of labourers...
Declan Butler, 'Are You Ready for the Revolution?', *Nature*, 2001, vol 409, pp 758–61
Page 249 – 'starts out with great promise, offering...'
Herbert Dreyfus, *What Computers Still Can't Do* (MIT Press, 1992)
Page 249 – 'Theories of evolution reinforce the value system...'
John Durant, 'The Myth of Human Evolution', *New University Quarterly*, 1981, vol 35, pp 425–38; see also Misia Landau, *Narratives of Human Education* (Yale University Press, 1991)
Page 250 – 'We eat well, we drink well...'
Michael Ignatieff, 'The Ascent of Man', *Prospect*, October 1999, pp 28–31
Page 250 – That optimism was underpinned...
Kenan Malik (Weidenfeld & Nicolson, 2000), op. cit.

Page 251 – 'On the maps provided. . .'
Bryan Appleyard, *Understanding the Present: Science and the Soul of Modern Man* (Picador, 1992)
Page 252 – It is only by recognising the narrow confines. . .
Roger Scruton, *An Intelligent Person's Guide to Philosophy* (Duckworth, 1997)

Chapter 10: Restoring Man to his Pedestal

Page 255 – The philosophic view alone recognises. . .
Thomas Kuhn, *The Structure of Scientific Revolutions* (Chicago University Press, 1970)
Page 255 – We are, it seems, on the brink. . .
R C Stroham, 'The Coming Revolution in Biology', *Nature: Biotechnology*, 1997, vol 15, pp 194–200
Page 256 – The major beneficiary of this new understanding. . .
John Maddox, *What Remains to be Discovered* (Macmillan, 1998); James Trefil, *1001 Things You Don't Know About Science – and No-One Else Does Either* (Cassel, 1996)
Page 256 – Biologists, freed from the obligation. . .
Robert Augros and George Stanciu, *The New Biology: Discovering the Wisdom in Nature* (New Science Library, 1988); see also Robert Wesson (MIT Press, 1993), op. cit.
Page 257 – A greater challenge still would be. . .
Rupert Sheldrake, *The Presence of the Past: Morphic Resonance and the Habits of Nature* (Park Street Press, 1988); Fritjof Capra (Flamingo, 1997), op. cit.; Scott F Gilbert et al., 'Resynthesizing Evolutionary and Developmental Biology', *Developmental Biology*, 1996, vol 137, pp 357–72; Michael J Denton et al., 'The Protein Fields as Platonic Forms: New Support for the Pre-Darwinian Conception of Evolution by Natural Law', *Journal of Theoretical Biology*, 2002, vol 219, pp 324–42; Jeffrey Schwartz et al., 'Quantum Physics in Neuroscience and Psychology: A Neurophysical Model of Mind-Brain Interaction', *Phil Trans R Soc B*, 2005, vol 360, pp 1309–27
Page 258 – The new paradigm must also lead to a. . .
John Polkinghorne, *Beyond Science: The Wider Human Context* (Cambridge University Press, 1995); see also Huston Smith, *Why Religion Matters* (Harper San Francisco, 2001); Keith Ward, *God, Chance and Necessity* (One World Publications, 1996); Denis Alexander, *Rebuilding the Matrix: Science and Faith in the Twenty-First Century* (Lion Publishing, 2001)
Page 258 – But while it is as hard as ever to imagine. . .
The most persuasive contemporary argument in favour of intelligent design is found in Michael J Behe, *Darwin's Black Box: The Biochemical Challenge to Evolution* (The Free Press, 1996)

Page 259 – This puts a rather different gloss on. . .
Mikael Stenmark, *Scientism: Science, Ethics and Religion* (Ashgate, 2001); see also Richard Dawkins, *The God Delusion* (Bantam Press, 2006); Lewis Wolpert, *Six Impossible Things Before Breakfast: The Evolutionary Origins of Belief* (Norton, 2007); Daniel Dennett, *Breaking the Spell: Religion as a Natural Phenomenon* (Penguin, 2007)
Page 261 – 'Darwin made theological. . .'
Douglas Futuyma, *Evolutionary Biology* (Sinnauer Assocs Inc, 1986)

Index

Page numbers in *italics* refer to illustrations